Erwin Dee Kord (Ed.)

Sene (Ghana Parliament Constituency)

Erwin Dee Kord (Ed.)

Sene (Ghana Parliament Constituency)

Tadeusz Sendzimir

Solv

Imprint

Publisher:
Solv is a trademark of
International Book Market Service Ltd., 17 Rue Meldrum, Beau Bassin, 1713-01 Mauritius
Email: info@bookmarketservice.com
Website: www.bookmarketservice.com

Published in 2011

Printed in: U.S.A., U.K., Germany. This book was not produced in Mauritius.

ISBN: 978-613-8-88667-9

Contents

Articles

References

Sendzimir_process

Sendzimir process (named after Tadeusz Sendzimir) is used to galvanize a steel strip by using a small amount of aluminum in the zinc bath and producing a coating with essentially no iron-zinc alloy. The process guarantees high resistance and durability characteristics. About 75% of hydrogen was needed in the original Sendzimir process but all the newer nonoxidizing methods of degreasing require only 7—15%.[1]

The rolling of hot steel slabs using a Sendzimir mill requires a much smaller operational area than a continuous hot strip mill. This milling process is not recommended for heavy duty running surfaces such as crane rail

See also

- Hot-dip galvanizing

References

[1] F. Porter. *Zinc Handbook.* 1991.

Tadeusz_Sendzimir

Tadeusz Sendzimir (originally **Sędzimir,**[1] July 15, 1894, Lwów – September 1, 1989, Jupiter, Florida[2]) of Ostoja coat of arms was a Polish engineer and inventor of international renown with 120 patents in mining and metallurgy, 73 of which were awarded to him in the United States.[3]

His name has been given to revolutionary methods of processing steel and metals used in every industrialized nation of the world. Sendzimir was a holder of the Polish Golden Cross of Merit (from 1938), the Bessemer Gold Medal (from 1965) and the Brinell Gold Medal from the Royal Swedish Academy of Sciences in Stockholm (from 1974). On the 100th anniversary of the Statue of Liberty he was one of those prominent immigrants honored for their contributions to America.

In 1990 Poland's large steel plant in Kraków (formerly the Lenin Steelworks) was renamed to Tadeusz Sendzimir Steelworks. The AIST Tadeusz Sendzimir Memorial Medal was established in the same year.

Early years

Sendzimir was the eldest son of Kazimierz Sędzimir belonging to the Clan of Ostoja and Wanda Jaskółowska. Fascinated by machinery as a child, he built his own camera at the age of 13. After studying at the 4th Classical Gymnasium (*Gimnazjum Klasyczne*) in Lwów he entered the Politechnika Lwowska. However when Lwów was captured by Russian troops the Polytechnic Institute was closed and he moved to Kiev. There he worked in auto services and in the Russian-American Chamber of commerce where he learned Russian and English.

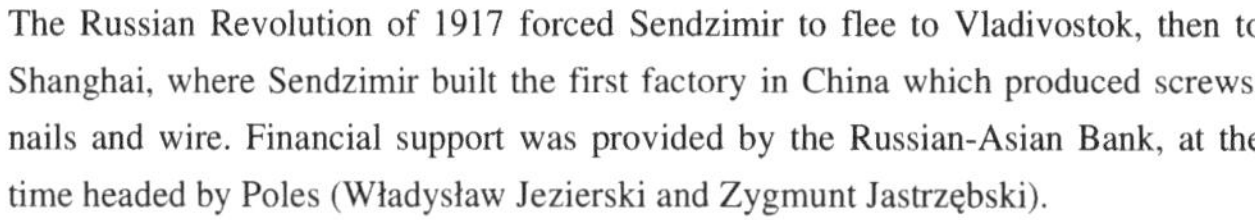

The Russian Revolution of 1917 forced Sendzimir to flee to Vladivostok, then to Shanghai, where Sendzimir built the first factory in China which produced screws, nails and wire. Financial support was provided by the Russian-Asian Bank, at the time headed by Poles (Władysław Jezierski and Zygmunt Jastrzębski).

Immigration and researches

In 1922 Sendzimir married Barbara Alferieff. His first son Michael was born two years later. Designing and making his own machines, Sendzimir began experimenting with a new way to galvanize steel. Despite galvanization, the products still had a tendency to oxidize. Sendzimir discovered that the problem was due to the zinc bonding to a thin layer of iron hydroxide on its surface, rather than to the iron.

In 1929 Sendzimir approached several American industrialists with his findings, but since the Great Depression had begun, he was unsuccessful in gaining their interest. Returning to Poland in 1931, he obtained support for the construction of the first industrial-scale galvanizing unit and put several cold strip mills into operation. This process was explained by him as follows: "Let's imagine a piece of a hard pastry. We are rolling it on the pastry-board to decrease its thickness. However it would be faster and easier if we asked someone to stretch it by holding the edges".

A steel mill in Butler, Pennsylvania was established by Sendzimir in 1936. By 1938 Armco Steel became interested in his work and they formed a partnership with him, the Armzen Company, to oversee the worldwide expansion of his galvanizing and mill technology. In the spring of 1939 Sendzimir moved to Middletown, Ohio. Sendzimir's patented mill could roll hard materials down to very light gauges. The U.S. company, T. Sendzimir, Inc., was formed by him in Waterbury, Connecticut, in the 1940s

In 1945, he married Bertha Bernoda. The following year he became a citizen of the United States. With the beginning of the Cold War, Sendzimir and his achievements were ignored by Communist Poland and he was not even mentioned in the Polish *Encyklopedia Powszechna* (Universal Encyclopedia). This changed when Edward Gierek, a new leader of the Polish United Workers' Party, came to power. Sendzimir was awarded an Officer Cross of the Order Polonia Restituta (*Krzyż Oficerski Orderu Odrodzenia Polski*).

In 1975 Sendzimir received the honorary degree of doctor *honoris causa* from the AGH University of Science and Technology in Kraków. Sendzimir's successful methods for galvanizing steel eventually were implemented in the first Z-mill rolling silicon steel, making it pliable enough for use in air defense radar. Between 1953 and 1989 he introduced the first productive Z-mill to Great Britain, and to Japan and Canada in the 1950s and 1960s. In 1974 Sendzimir invented a spiral steel looper used in both the United States and Japan.

Ninety percent of the world's stainless steel production went through the Sendzimir process by the early 1980s. Poland, France, the United Kingdom, Japan, and Canada purchased his steel mills and technologies over the years. Most notably, Sendzimir was a major financial and personal supporter of the Kościuszko Foundation, the Polish Institute of Arts and Sciences of America and Alliance College in Pennsylvania. Sendzimir died after a massive stroke and was buried by his family in a zinc-plated coffin made according to his design.

Quote

- "I have been carrying family and the only my capital - a new method of zinc-plating to another coast of the Pacific"

Notes

[1] As of the 17th century version. The surname was changed after the second arrival to the United States.

[2] According to *Polski Słownik Biograficzny* he died in Jupiter, Florida.

[3] Sarmatian Review XV.1 (http://www.ruf.rice.edu/~sarmatia/195/mitchell.html).

See also

- Sendzimir process
- Sendzimir planetary mill
- Berthe Sendzimir
- Ostoja coat of arms
- Clan of Ostoja

References

- Vanda Sendzimir. *Steel Will: The Life of Tad Sendzimir.* New York, Hippocrene Books, 1994.
- M. Kalisz. *Walcownia znaczy Sendzimir.* "Przekrój", 1973, nr. 1468.
- O. Budrewicz. *Ocynkowane życie.* "Perspektywy", 1974, nr. 38.

External links

- The Sendzimir Foundation site (http://www.sendzimir.org.pl/)
- Sendzimir mills (http://www.isherman.com/Inventory/SZM-_Sendzimir_Mills/szm-_sendzimir_mills.htm)
- Tadeusz Gajl, Herbarz Polski, Sędzimir CoA (http://herbyszlachty.pl/herby_user_herb.php?lang=pl&herb=sedzimir)

]]

Galvanization

Galvanization (or **galvanisation**) is the process of applying a protective zinc coating to steel or iron, in order to prevent rusting. The term is derived from the name of Italian scientist Luigi Galvani.

Although galvanization can be done with electrochemical and electrodeposition processes, the most common method in current use is hot-dip galvanization, in which steel parts are submerged in a bath of molten zinc.

Meanings

Metal protection

In current use, the term refers to the coating of steel or iron with zinc. This is done to prevent rusting of the ferrous item. The value of galvanizing stems from the relative corrosion resistance of zinc, which, under most service conditions, is considerably less than those of iron and steel. The zinc therefore serves as a sacrificial anode, so that it cathodically protects exposed steel. This means that even if the coating is scratched or abraded, the exposed steel will still be protected from corrosion by the remaining zinc - an advantage absent from paint, enamel, powder coating and other methods. Galvanizing is also favored as a means of protective coating because of its low cost, ease of application and comparatively long maintenance-free service life.

The term galvanizing, while technically referring specifically to the application of zinc coating by the use of a galvanic cell (also known as electroplating), is also generally understood to include hot-dip zinc coating. The practical difference is that hot-dip galvanization produces a thick, durable and matte gray coating - electroplated coatings tend to be thin and brightly reflective. Due to its thinness, the zinc of electroplated coatings is quickly depleted, making them unsuitable for outdoor applications (except in very dry climates). When combined with subsequent painting (which slows zinc consumption), electroplating is durable enough to be used in some premium auto body coatings.

Nonetheless, electroplating is used on its own for many outdoor applications because it is cheaper than hot dip zinc coating and looks good when new. Another reason not to use hot dip zinc coating is that for bolts and nuts size M10 or smaller, the thick hot-dipped coating uses up too much of the threads, which reduces strength (because the dimension of the steel prior to coating must be reduced for the fasteners to fit together). This means that for cars, bicycles and many other 'light' mechanical products, the alternative to electroplating bolts and nuts is not hot dip zinc coating but making the bolts and nuts from stainless steel (known by the corrosion grades A4 and A2).

Electroplated steel is visually indistinguishable from stainless steel when new. To determine whether a part is electroplated or stainless steel, apply a magnet. The most common stainless steel alloys (including those used for bolts and nuts) are not magnetic or only very slightly attracted to a magnet.

History

Originally, "galvanization" was the administration of electric shocks (in the 19th century also termed *Faradism*, after Michael Faraday). It stemmed from Galvani's induction of twitches in severed frogs' legs, by his accidental generation of electricity. Its claims to health benefits have largely been disproved, except for some limited uses in psychiatry in the form of electroconvulsive therapy (ECT). This archaic sense is the origin of the meaning of *galvanic* when meaning "affected/affecting, as if by a shock of electricity; startled".[1] and the metaphorical "galvanize into action" referring to suddenly stimulating a complacent person or group to take action. Later the word was used for processes of electrodeposition. This remains a useful and broadly applied technology, but the term "galvanization" has largely come to be associated with zinc coatings, to the exclusion of other metals.

Galvanic paint, a precursor to hot-dip galvanization, was patented by Stanislas Sorel, of Paris, France in December, 1837.[2]

The earliest known example of galvanizing of iron was found on the 30th September 1999 by the Royal Armouries Museum on a 17th century Indian armour in their collection.[3]

Zinc coatings

Zinc coatings prevent corrosion of the protected metal by forming a physical barrier, and by acting as a sacrificial anode if this barrier is damaged. When exposed to the atmosphere, zinc reacts with oxygen to form zinc oxide, which further reacts with water molecules in the air to form zinc hydroxide. Finally zinc hydroxide reacts with carbon dioxide in the atmosphere to yield a thin, impermeable, tenacious and quite insoluble dull gray layer of zinc carbonate which adheres extremely well to the underlying zinc, so protecting it from further corrosion, in a way similar to the protection afforded to aluminium and stainless steels by their oxide layers.

Hot-dip galvanizing deposits a thick robust layer that may be more than is necessary for the protection of the underlying metal in some applications. This is the case in automobile bodies, where additional rust proofing paint will be applied. Here, a thinner form of galvanizing is applied by electroplating, called "electrogalvanization". The hot-dip process slightly reduces the strength of the base metal, which is a consideration for the manufacture of wire rope and other highly-stressed products. The protection provided by this process is insufficient for products that will be constantly exposed to corrosive materials such as salt water. For these applications, more expensive stainless steel is preferred. Some nails made today are electro-galvanized.

As noted previously, both mechanisms are often at work in practical applications. For example, the traditional measure of a coating's effectiveness is resistance to a salt spray. Thin coatings cannot remain intact indefinitely when subject to surface abrasion, and the galvanic protection offered by zinc can be sharply contrasted to more noble metals. As an example, a scratched or incomplete coating of chromium actually exacerbates corrosion of the underlying steel, since it is less electrochemically active than the substrate.

The size of crystallites in galvanized coatings is an aesthetic feature, known as **spangle**. By varying the number of particles added for heterogeneous nucleation and the rate of cooling in a hot-dip process, the spangle can be adjusted from an apparently uniform surface (crystallites too small to see with the naked eye) to grains several centimetres wide. Visible crystallites are rare in other engineering materials. Protective coatings for steel constitute the largest use of zinc and rely upon the galvanic or sacrificial property of zinc relative to steel.

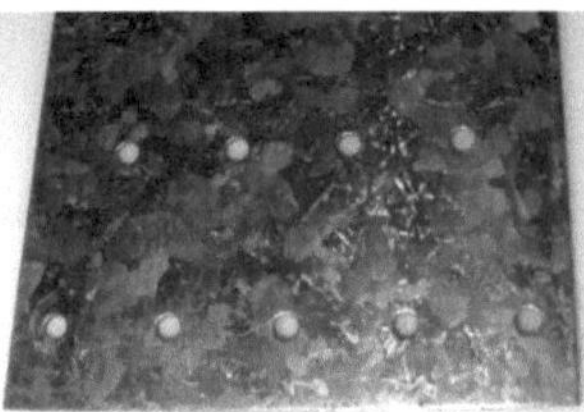

Galvanized surface with visible spangle

Thermal diffusion galvanizing, a form of Sherardizing, provides a zinc coating on iron or copper based materials partially similar to hot dip galvanizing. The final surface is different than hot-dip Galvanizing; all of its zinc is alloyed.[4] Zinc is applied in a powder form with "accelerator chemicals" (generally sand,[5] but other chemicals are patented). The parts and the zinc powder are tumbled in a sealed drum while it is heated to slightly below zinc's melting temperature. The drum must be heated evenly, or complications will arise. Due to the chemicals added to the zinc powder, the zinc/iron makes an alloy at a lower temperature than hot dip galvanizing. This process requires generally fewer preparatory cleanings than other methods. The dull-grey crystal structure formed by the process bonds stronger with paint, powder coating, and rubber overmolding processes than other methods. It is a preferred method for coating small, complex-shaped metals and smoothing in rough surfaces on items formed with powder metal.

Eventual corrosion

Rusted corrugated steel roof

Although galvanizing will inhibit attack of the underlying steel, rusting will be inevitable, especially due to natural acidity of rain. For example, corrugated iron sheet roofing will start to degrade within a few years despite the protective action of the zinc coating. Marine and salty environments also lower the lifetime of galvanized iron because the high electrical conductivity of sea water increases the rate of corrosion. Galvanized car frames exemplify this; they corrode much quicker in cold environments due to road salt. Galvanized steel can last for many years if other means are maintained, such as paint coatings and additional sacrificial anodes.

See also

- Corrugated galvanised iron
- Galvannealed - galvanization and annealing
- Sendzimir process
- Sherardising

References

[1] "Galvanic; Dictionary.com" (http://dictionary.reference.com/search?r=10&q=galvanic). . Retrieved 2006-11-30.

[2] Process for protecting articles made of Iron or Steel from oxidation." (http://books.google.co.uk/books?id=W8oGAAAAYAAJ&pg=PA52&dq="Specification+of+a+Patent+for+a+process+for+protecting+articles+made+of+Iron+or+Steel+from+oxidation."&lr=&as_brr=0&as_pt=ALLTYPES) Specification of patent granted to M. Sorel, of Paris, France, December, 1837. Journal of the Franklin Institute (Philadelphia, Pa.), Published by Pergamon Press, 1838, via Google Book Search.

[3] (http://www.royalarmouries.org/what-we-do/research/analytical-projects/zinc-coatings-of-indian-plate-and-mail-armour) Summary of XRF analysis conducted at the Royal Armouries Museum in Leeds and written up as part of a thesis by Helen Bowstead Stallybrass at the Department of Archaeological Sciences, Bradford University.

[4] Presentation on Thermal Diffusion Galvanizing: http://www.armycorrosion.com/past_summits/summit2009/09Presentations%5CDay3%5CMosheMoked.pdf

[5] Porter, Frank C. (1991). *Zinc Handbook* (http://books.google.com/books?id=laACw9i0D_wC&q=sherardizing&cad=6#v=snippet&q=sherardizing&f=false). CRC Press. ISBN 9780824783402. .

External links

- Life Program regarding liquid waste management in galvanic processes. (http://www.zeroplus.eu:http://ec.europa.eu/environment/life/)
- American Galvanizers Association (http://www.galvanizeit.org/)

Strip_steel

Strip Steel or **cold rolled strip** is a steel product that is produced from a hot rolled strip that has been pickled. The coil is then reduced by a single stand cold roll steel mill straight away or reversing mill or in a tandem mill consisting of several single stands in a series. The strip is reduced to approximately final thickness by cold-rolling directly, or with the inclusion of an annealing operation at some intermediate thickness to facilitate further cold reduction or to obtain mechanical properties desired in the finished product. High carbon strip steel requires additional annealing and cold reduction operations[1] . The coil is then slit to the desired width through the process of roll slitting.

The final product typically consists of cold rolled steel that has been cut into strips of a specific widths and coiled or oscillate coiled for delivery, frequently interleaved with paper or another material which protects the surface finish of the material and assists in retaining oil or some other rust prevention solution. This product is often later stamped to form a part from the strip steel.

References

[1] "Glossary at Gibraltar Strip Steel" (http://www.worthingtonsteel.com/helpfulinfo/glossary_ab.php) Retrieved on 2009-08-27

Redox

Redox (portmanteau for ***red*uction-*ox*idation**) reactions describe all chemical reactions in which atoms have their oxidation state changed. This can be either a simple redox process, such as the oxidation of carbon to yield carbon dioxide (CO_2) or the reduction of carbon by hydrogen to yield methane (CH_4), or a complex process such as the oxidation of glucose ($C_6H_{12}O_6$) in the human body through a series of complex electron transfer processes.

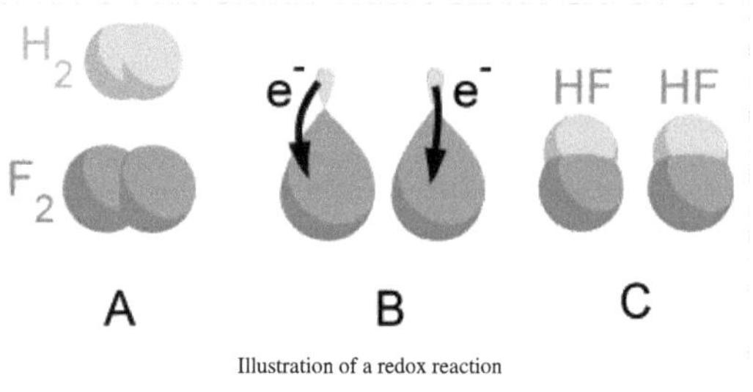

Illustration of a redox reaction

Redox reactions, or oxidation-reduction reactions, have a number of similarities to acid-base reactions. Fundamentally, redox reactions are a family of reactions that are concerned with the transfer of electrons between species. Like acid-base reactions, redox reactions are a matched set, that is there cannot be an oxidation reaction without a reduction reaction happening simultaneously. Oxidation refers to the loss of electrons, while reduction refers to the gain of electrons. Each reaction by itself is called a "half-reaction", simply because there must be two half-reactions to form a whole reaction. Thus, in notating redox reactions, chemists typically write out the electrons explicitly:

The term comes from the two concepts of reduction and oxidation. It can be explained in simple terms:

- **Oxidation** is the *loss* of electrons or an *increase* in oxidation state by a molecule, atom, or ion.
- **Reduction** is the *gain* of electrons or a *decrease* in oxidation state by a molecule, atom, or ion.

Though sufficient for many purposes, these descriptions are not precisely correct. Oxidation and reduction properly refer to *a change in oxidation state* — the actual transfer of electrons may never occur. Thus, oxidation is better defined as an *increase in oxidation state*, and reduction as a *decrease in oxidation state*. In practice, the transfer of electrons will always cause a change in oxidation state, but there are many reactions that are classed as "redox" even though no electron transfer occurs (such as those involving covalent bonds).

Non-redox reactions, which do not involve changes in formal charge, are known as metathesis reactions.

Oxidizing and reducing agents

Reduction

Oxidant + e^- ⟶ Product

(Electrons **gained**; oxidation number **decreases**)

Oxidation

Reductant ⟶ Product + e^-

(Electrons **lost**; oxidation number **increases**)

The two parts of a redox reaction

In redox processes, the reductant transfers electrons to the oxidant. Thus, in the reaction, the reductant or *reducing agent* loses electrons and is oxidized, and the oxidant or *oxidizing agent* gains electrons and is reduced. The pair of an oxidizing and reducing agent that are involved in a particular reaction is called a **redox pair**.

Oxidizers

Rusting iron

Substances that have the ability to **oxidize** other substances are said to be **oxidative** or **oxidizing** and are known as oxidizing agents, **oxidants**, or **oxidizers**. Put another way, the oxidant (oxidizing agent) removes electrons from another substance; i.e., it oxidizes other substances, and is thus itself reduced. And, because it "accepts" electrons, it is also called an electron acceptor.

Oxidants are usually chemical elements or substances with elements in high oxidation states (e.g., H_2O_2, MnO, CrO_3, Cr_2O, OsO_4) or highly electronegative substances/elements that can gain one or two extra electrons by oxidizing an element or substance (O, F, Cl, Br).

Reducers

A bonfire. Combustion consists of redox reactions involving free radicals.

Substances that have the ability to **reduce** other substances are said to be **reductive** or **reducing** and are known as reducing agents, **reductants**, or **reducers**. The reductant (reducing agent) transfers electrons to another substance; i.e., it reduces others, and is thus itself oxidized. And, because it "donates" electrons, it is also called an electron donor. Electron donors can also form charge transfer complexes with electron acceptors.

Reductants in chemistry are very diverse. Electropositive elemental metals, such as lithium, sodium, magnesium, iron, zinc, and aluminium, are good reducing agents. These metals donate or *give away* electrons readily. *Hydride transfer reagents*, such as $NaBH_4$ and $LiAlH_4$, are widely used in organic chemistry,[1] [2] primarily in the reduction of carbonyl compounds to alcohols. Another method of reduction involves the use of hydrogen gas (H_2) with a palladium, platinum, or nickel catalyst. These *catalytic reductions* are used primarily in the reduction of carbon-carbon double or triple bonds.

Standard reduction potential

Reduction potential is used to calculate the standard electrode potential (E^ocell).

This is the equation most commonly seen in textbooks:

$\mathbf{E^o cell = E^o_{red} + E^o_{ox}}$.

where: E^ocell is the standard electrode potential (in volts).

E^o_{red} is standard reduction potential of the reducing agent.

E^o_{ox} (standard oxidation potential) is negative of the standard reduction potential of the oxidizing agent.

though the following equation is generally more useful as one is usually given only reduction potentials, not oxidation potentials:

$\mathbf{E^o cell = E^o_{red} - E^o_{ox}}$

or the equivalent:

$\mathbf{E^o cell = E^o_{cathode} - E^o_{anode}}$

where:

E^ocell is the standard electrode potential (in volts).

E^o_{red} ($E^o_{cathode}$) is standard reduction potential of the reducing agent.

E^o_{ox} (E^o_{anode}) is the standard reduction potential of the oxidizing agent.

Examples of redox reactions

A good example is the reaction between hydrogen and fluorine in which hydrogen is being oxidized and fluorine is being reduced:

$$H_2 + F_2 \rightarrow 2\ HF$$

We can write this overall reaction as two half-reactions:

the oxidation reaction:

$$H_2 \rightarrow 2\ H^+ + 2\ e^-$$

and the reduction reaction:

$$F_2 + 2\ e^- \rightarrow 2\ F^-$$

Analyzing each half-reaction in isolation can often make the overall chemical process clearer. Because there is no net change in charge during a redox reaction, the number of electrons in excess in the oxidation reaction must equal the number consumed by the reduction reaction (as shown above).

Elements, even in molecular form, always have an oxidation state of zero. In the first half-reaction, hydrogen is oxidized from an oxidation state of zero to an oxidation state of +1. In the second half-reaction, fluorine is reduced from an oxidation state of zero to an oxidation state of −1.

When adding the reactions together the electrons are canceled:

$$H_2 \rightarrow 2\,H^+ + 2\,e^-$$

$$F_2 + 2\,e^- \rightarrow 2\,F^-$$

$$H_2 + F_2 \rightarrow 2\,H^+ + 2\,F^-$$

And the ions combine to form hydrogen fluoride:

$$H_2 + F_2 \rightarrow 2\,H^+ + 2\,F^- \rightarrow 2\,HF$$

Displacement reactions

Redox occurs in single displacement reactions or substitution reactions. The redox component of these types of reactions is the change of oxidation state (charge) on certain atoms, not the actual exchange of atoms in the compounds.

For example, in the reaction between iron and copper(II) sulfate solution:

$$Fe + CuSO_4 \rightarrow FeSO_4 + Cu$$

The ionic equation for this reaction is:

$$Fe + Cu^{2+} \rightarrow Fe^{2+} + Cu$$

As two half-equations, it is seen that the iron is oxidized:

$$Fe \rightarrow Fe^{2+} + 2\,e^-$$

And the copper is reduced:

$$Cu^{2+} + 2\,e^- \rightarrow Cu$$

Other examples

- The oxidation of iron(II) to iron(III) by hydrogen peroxide in the presence of an acid:

  $$Fe^{2+} \rightarrow Fe^{3+} + e^-$$

  $$H_2O_2 + 2\,e^- \rightarrow 2\,OH^-$$

  Overall equation:

  $$2\,Fe^{2+} + H_2O_2 + 2\,H^+ \rightarrow 2\,Fe^{3+} + 2\,H_2O$$

- The reduction of nitrate to nitrogen in the presence of an acid (denitrification):

  $$2\,NO_3^- + 10\,e^- + 12\,H^+ \rightarrow N_2 + 6\,H_2O$$

- Oxidation of elemental iron to iron(III) oxide by oxygen (commonly known as rusting, which is similar to tarnishing):

  $$4\,Fe + 3\,O_2 \rightarrow 2\,Fe_2O_3$$

- The combustion of hydrocarbons, such as in an internal combustion engine, which produces water, carbon dioxide, some partially oxidized forms such as carbon monoxide, and heat energy. Complete oxidation of materials containing carbon produces carbon dioxide.
- In organic chemistry, the stepwise oxidation of a hydrocarbon by oxygen produces water and, successively, an alcohol, an aldehyde or a ketone, a carboxylic acid, and then a peroxide.

Iron rusting in pyrite cubes

Redox reactions in industry

The primary process of reducing ore to produce metals is discussed in the article on Smelting.

Oxidation is used in a wide variety of industries such as in the production of cleaning products and oxidizing ammonia to produce nitric acid, which is used in most fertilizers.

Redox reactions are the foundation of electrochemical cells.

The process of electroplating uses redox reactions to coat objects with a thin layer of a material, as in chrome-plated automotive parts, silver plating cutlery, and gold-plated jewelry.

The production of compact discs depends on a redox reaction, which coats the disc with a thin layer of metal film.

Redox reactions in biology

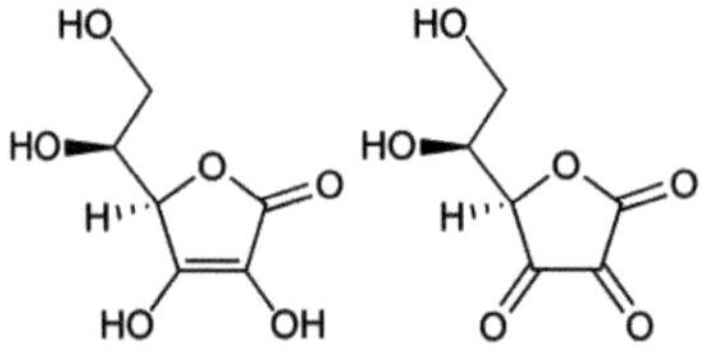

Top: ascorbic acid (reduced form of Vitamin C)

Bottom: dehydroascorbic acid (oxidized form of Vitamin C)

Many important biological processes involve redox reactions.

Cellular respiration, for instance, is the oxidation of glucose ($C_6H_{12}O_6$) to CO_2 and the reduction of oxygen to water. The summary equation for cell respiration is:

$$C_6H_{12}O_6 + 6\ O_2 \rightarrow 6\ CO_2 + 6\ H_2O$$

The process of cell respiration also depends heavily on the reduction of NAD^+ to NADH and the reverse reaction (the oxidation of NADH to NAD^+). Photosynthesis and Cellular respiration are complementary but photosynthesis is not the reverse of the redox reaction in cell respiration:

$$6\ CO_2 + 6\ H_2O + \text{light energy} \rightarrow C_6H_{12}O_6 + 6\ O_2$$

Biological energy is frequently stored and released by means of redox reactions. Photosynthesis involves the reduction of carbon dioxide into sugars and the oxidation of water into molecular oxygen. The reverse reaction, respiration, oxidizes sugars to produce carbon dioxide and water. As intermediate steps, the reduced carbon compounds are used to reduce nicotinamide adenine dinucleotide (NAD^+), which then contributes to the creation of a proton gradient, which drives the synthesis of adenosine triphosphate (ATP) and is maintained by the reduction of oxygen. In animal cells, mitochondria perform similar functions. See *Membrane potential* article.

Free radical reactions are redox reactions that occur as a part of homeostasis and killing microorganisms, where an electron detaches from a molecule and then reattaches almost instantaneously. Free radicals are a part of redox molecules and can become harmful to the human body if they do not reattach to the redox molecule or an antioxidant. Unsatisfied free radicals can spur the mutation of cells they encounter and are thus causes of cancer.

The term **redox state** is often used to describe the balance of NAD^+/NADH and $NADP^+$/NADPH in a biological system such as a cell or organ. The redox state is reflected in the balance of several sets of metabolites (e.g., lactate and pyruvate, beta-hydroxybutyrate and acetoacetate), whose interconversion is dependent on these ratios. An abnormal redox state can develop in a variety of deleterious situations, such as hypoxia, shock, and sepsis. Redox signaling involves the control of cellular processes by redox processes.

Redox proteins and their genes must be co-located for redox regulation according to the CoRR hypothesis for the function of DNA in mitochondria and chloroplasts.

Redox cycling

A wide variety of aromatic compounds are enzymatically reduced to form free radicals that contain one more electron than their parent compounds. In general, the electron donor is any of a wide variety of flavoenzymes and their coenzymes. Once formed, these anion free radicals reduce molecular oxygen to superoxide, and regenerate the unchanged parent compound. The net reaction is the oxidation of the flavoenzyme's coenzymes and the reduction of molecular oxygen to form superoxide. This catalytic behavior has been described as futile cycle or redox cycling.

Examples of redox cycling-inducing molecules are the herbicide paraquat and other viologens and quinones such as menadione.[3]

Redox reactions in geology

In geology, redox is important to both the formation of minerals, mobilization of minerals, and in some depositional environments. In general, the redox state of most rocks can be seen in the color of the rock. Red is associated with oxidizing conditions of formation, and green is typically associated with reducing conditions. White (bleached rock) can also be associated with reducing conditions. Famous examples of redox conditions affecting geological processes include uranium deposits and Moqui marbles.

A uranium mine, near Moab, Utah. Note alternating red and white/green sandstone. This corresponds to oxidized and reduced conditions in groundwater redox chemistry. The rock forms in oxidizing conditions, and is then "bleached" to the white/green state when a reducing fluid passes through the rock. The reduced fluid can also carry uranium-bearing minerals.

Balancing redox reactions

Describing the overall electrochemical reaction for a redox process requires a *balancing* of the component half-reactions for oxidation and reduction. In general, for reactions in aqueous solution, this involves adding H^+, OH^-, H_2O, and electrons to compensate for the oxidation changes.

Acidic media

In acidic media, H^+ ions and water are added to half reactions to balance the overall reaction.

For example, when manganese(II) reacts with sodium bismuthate:

Unbalanced reaction:	$Mn^{2+}(aq) + NaBiO_3(s) \rightarrow Bi^{3+}(aq) + MnO_4^-(aq)$
Oxidation:	$4\,H_2O(l) + Mn^{2+}(aq) \rightarrow MnO(aq) + 8\,H^+(aq) + 5\,e^-$
Reduction:	$2\,e^- + 6\,H^+ + BiO(s) \rightarrow Bi^{3+}(aq) + 3\,H_2O(l)$

The reaction is balanced by scaling the two half-cell reactions to involve the same number of electrons (multiplying the oxidation reaction by the number of electrons in the reduction step and vice versa):

$$8\,H_2O(l) + 2\,Mn^{2+}(aq) \rightarrow 2\,MnO(aq) + 16\,H^+(aq) + 10\,e^-$$

$$10\,e^- + 30\,H^+ + 5\,BiO(s) \rightarrow 5\,Bi^{3+}(aq) + 15\,H_2O(l)$$

Adding these two reactions eliminates the electrons terms and yields the balanced reaction:

$$14\,H^+(aq) + 2\,Mn^{2+}(aq) + 5\,NaBiO_3(s) \rightarrow 7\,H_2O(l) + 2\,MnO(aq) + 5\,Bi^{3+}(aq) + 5\,Na^+(aq)$$

Basic media

In basic media, OH^- ions and water are added to half reactions to balance the overall reaction.

For example, in the reaction between potassium permanganate and sodium sulfite:

Unbalanced reaction: $KMnO_4 + Na_2SO_3 + H_2O \rightarrow MnO_2 + Na_2SO_4 + KOH$

Reduction: $3\ e^- + 2\ H_2O + MnO_4^- \rightarrow MnO_2 + 4\ OH^-$

Oxidation: $2\ OH^- + SO_3^{2-} \rightarrow SO_4^{2-} + H_2O + 2\ e^-$

Balancing the number of electrons in the two half-cell reactions gives:

$$6\ e^- + 4\ H_2O + 2\ MnO_4^- \rightarrow 2\ MnO_2 + 8\ OH^-$$

$$6\ OH^- + 3\ SO_3^{2-} \rightarrow 3\ SO_4^{2-} + 3\ H_2O + 6\ e^-$$

Adding these two half-cell reactions together gives the balanced equation:

$$2\ KMnO_4 + 3\ Na_2SO_3 + H_2O \rightarrow 2\ MnO_2 + 3\ Na_2SO_4 + 2\ KOH$$

Memory aids

The key terms involved in redox are often confusing to students.[4] [5] For example, an element that is oxidized loses electrons; however, that element is referred to as the reducing agent. Likewise, an element that is reduced gains electrons and is referred to as the oxidizing agent.[6] Acronyms or mnemonics are commonly used[7] to help remember what is happening:

- "LEO the lion says GER" — **L**oss of **E**lectrons is **O**xidation, **G**ain of **E**lectrons is **R**eduction.[4] [5] [7] [6]
- "LEORA says GEROA" — **L**oss of **E**lectrons is **O**xidation (**R**educing **A**gent) and **G**ain of **E**lectrons is **R**educed (**O**xidizing **A**gent).[6]
- "OIL RIG"—**O**xidation **I**s **L**oss of electrons, **R**eduction **I**s **G**ain of electrons.[4] [5] [7] [6]

See also

- Bessemer process
- Bioremediation
- Calvin cycle
- Chemical equation
- Chemical looping combustion
- Citric acid cycle
- Electrochemical cell
- Electrochemical series
- Electrochemistry
- Electrolysis
- Electron equivalent
- Electron transport chain
- Galvanic cell
- Hydrogenation
- Membrane potential
- Organic redox reaction
- Oxidative addition and reductive elimination
- Oxidative phosphorylation
- Partial oxidation
- Pro-oxidant
- Reduced gas
- Reducing agent
- Reduction potential
- Thermic reaction

References

[1] Hudlický, Miloš (1996). *Reductions in Organic Chemistry*. Washington, D.C.: American Chemical Society. pp. 429. ISBN 0-8412-3344-6.

[2] Hudlický, Miloš (1990). *Oxidations in Organic Chemistry*. Washington, D.C.: American Chemical Society. pp. 456. ISBN 0-8412-1780-7.

[3] "gutier.doc" (http://www.bioscience.org/2000/v5/d/gutier/gutier.pdf). . Retrieved 2008-06-30.PDF (2.76 MiB)

[4] Robertson, William (2010). *More Chemistry Basics* (http://books.google.com/books?id=hIzuarlXtH4C&pg=PA82&dq=confusing). National Science Teachers Association. p. 82. ISBN 978-1-936137-74-9. .

[5] Phillips, John; Strozak, Victor; Wistrom, Cheryl (2000). *Chemistry: Concepts and Applications*. Glencoe McGraw-Hill. p. 558. ISBN 978-0028282107. "Students often are confused when associating reduction with the gain of electrons."

[6] Rodgers, Glen (2012). *Descriptive Inorganic, Coordination, and Solid-State Chemistry* (http://books.google.com/books?id=g_ybia0hGw8C&pg=PA330). Brooks/Cole, Cengage Learning. p. 330. ISBN 978-0-8400-6846-0. .

[7] Zumdahl, Steven; Zumdahl, Susan (2009). *Chemistry* (http://books.google.com/books?id=IdhYqXy37KIC&pg=PA160). Houghton Mifflin. p. 160. ISBN 978-0-547-05405-6. .

- Schüring, J., Schulz, H. D., Fischer, W. R., Böttcher, J., Duijnisveld, W. H. (editors)(1999). Redox: Fundamentals, Processes and Applications, Springer-Verlag, Heidelberg, 246 pp. ISBN 978-3-540-66528-1 (pdf 3,6 MB) (http://hdl.handle.net/10013/epic.31694.d001)

External links

- Chemical Equation Balancer (http://www.berkeleychurchill.com/software/chembal.php) - An open source chemical equation balancer that handles redox reactions.
- Video - Synthesis of Copper(II) Acetate (http://www.youtube.com/watch?v=rF1ls-v7puQ) 20 Feb. 2009
- Redox reactions calculator (http://www.shodor.org/UNChem/advanced/redox/redoxcalc.html)
- Redox reactions at Chemguide (http://www.chemguide.co.uk/inorganic/redox/definitions.html#top)
- Online redox reaction equation balancer, balances equations of any half-cell and full reactions (http://www.webqc.org/balance.php)

Hot-dip_galvanizing

Hot-dip galvanizing is a form of galvanization. It is the process of coating iron, steel, or aluminum with a thin zinc layer, by passing the metal through a molten bath of zinc at a temperature of around 860 °F (460 °C). When exposed to the atmosphere, the pure zinc (Zn) reacts with *oxygen* (O_2) to form *zinc oxide* (ZnO), which further reacts with *carbon dioxide* (CO_2) to form *zinc carbonate* ($ZnCO_3$), a usually dull grey, fairly strong material that stops further corrosion in many circumstances, protecting the steel below from the elements. **Galvanized steel** is widely used in applications where rust resistance is needed, and can be identified by the crystallization patterning on the surface (often called a "spangle").

Crystalline surface of a hot-dip galvanized handrail

Process

The process of hot-dip galvanizing results in a metallurgical bond between zinc and steel with a series of distinct iron-zinc alloys. The resulting coated steel can be used in much the same way as uncoated. Galvanized steel can be welded; however, one must exercise caution around the resulting zinc fumes. Galvanized steel is suitable for high-temperature applications of up to 392°F (200°C). The use of galvanized steel at temperatures above this will result in peeling of the zinc at the *intermetallic* layer. Electrogalvanized sheet steel is often used in automotive manufacturing to enhance the corrosion performance of exterior body panels, this is however a completely different process.

Lead is often added to the molten zinc bath to improve the fluidity of the bath (thus limiting excess zinc on the dipped product by improved drainage properties), helps prevent floating dross, makes dross recycling easier and protects the kettle from uneven heat distribution from the burners.[1] Lead is either added to primary Z1 Grade Zinc or already contained in used secondary zinc. A third, declining method is to use low Z5 Grade Zinc.[2]

Steel strip can be hot-dip galvanized in a continuous line. Hot-dip galvanized steel strip (also sometimes loosely referred to as galvanized iron) is extensively used for applications requiring the strength of steel combined with the resistance to corrosion of zinc. Applications include: roofing and walling, safety barriers, handrails, consumer appliances and automotive body parts. One common use is in metal pails. Galvanised steel is also used in most heating and cooling duct systems in buildings

Individual metal articles, such as steel girders or wrought iron gates, can be hot-dip galvanized by a process called batch galvanizing. Other modern techniques have largely replaced hot-dip for these sorts of roles. This includes *electrogalvanizing*, which deposits the layer of zinc from an aqueous electrolyte by electroplating, forming a thinner and much stronger bond.

History

In 1742, French chemist Paul Jacques Malouin described a method of coating iron by dipping it in molten zinc in a presentation to the French Royal Academy. In 1836, French chemist Stanislas Sorel obtained a patent for a method of coating iron with zinc, after first cleaning it with 9% *sulfuric acid* (H_2SO_4) and fluxing it with *ammonium chloride* (NH_4Cl).

Media

A hot-dip galvanizing 'kettle' with fume hood

Galvanized I-beam

Galvanized angle iron

Steel tube ready to be galvanized

See also

- Electrogalvanization
- Plating
- Bolt manufacturing process
- Corrugated galvanised iron
- Galvannealed - galvanization and annealing
- Liquid metal embrittlement
- Metal fume fever
- Sendzimir process
- Surface finishing
- Thermal spraying

References

[1] Lohse et al., Adaptation to Scientific and Technical Progress of Annex II Directive 2000/53/EC, January 2008 (http://rohs-elv.exemptions.oeko.info/fileadmin/user_upload/Background/Final_report_ELV_Annex_II_revision.pdf)
Lead addition described on page 12

[2] European Commission, Reference Document on Best Available Techniques in the Non Ferrous Metals Industries, December 2001 (ftp://ftp.jrc.es/pub/eippcb/doc/nfm_bref_1201.pdf)
Primary zinc grade table on page 15

Institute_of_Materials,_Minerals_and_Mining

Institute of Materials, Minerals and Mining	
IOM3 The Institute of Materials, Minerals and Mining	
Abbreviation	IOM3
Motto	The global network for the materials cycle
Formation	2002
Legal status	Non-profit organization
Purpose/focus	Materials in the UK
Location	1 Carlton House Terrace, London, SW1Y 5DB
Region served	UK
Membership	UK materials scientists
Chief Executive	Dr Bernie Rickinson
Main organ	IOM3 Council
Website	IOM3 [1]

The **Institute of Materials, Minerals and Mining** (IOM3) is a major UK engineering institution whose activities encompass the whole materials cycle, from exploration and extraction, through characterisation, processing, forming, finishing and application, to product recycling and land reuse. It exists to promote and develop all aspects of materials science and engineering, geology, mining and associated technologies, mineral and petroleum engineering and extraction metallurgy, as a leading authority in the worldwide materials and mining community.

It is a registered charity governed by Royal Charter and in 2009-10 had a gross income of £6.4million.[2]

Structure

Its headquarters are at Carlton House Terrace off Pall Mall in St James's in central London.[3] The organisation also has its education, marketing and knowledge transfer office in Grantham, and its finance office Stoke on Trent.

Members of the Institute come from a variety of backgrounds, from students to company chief executives. Members qualify for different grades of membership, ranging from Affiliate to Fellow (FIMMM), depending on academic qualifications and professional experience. IOM3 has an individual membership of over 20,000, and represents a combination of scientific, technical and human resources which links industry, government, research and the academic world.

Around 40 UK 'local societies' are affiliated with the Institute, covering a wide range of disciplines including ceramics, composites, mining, packaging, polymers and metallurgy, and organising events throughout the year.

Divisions

- Applied Earth Science [4] - part of IMMa
- Automotive [5]
- Biomedical Applications [6]
- British Composites Society [7] - composite materials
- Casting [8] - for the castings industry
- Ceramics Society [9] - building materials, ceramics, cementitious materials, clay technology, refactories, and whitewares
- Electronics Applications [10]
- Energy Materials Group [11]
- IMMa: International Mining & Minerals Association [12]
- Iron & Steel Society [13]
- IWSc: The Wood Technology Society [14]
- Light Metals [15]
- Materials Science & Technology [16]
- Minerals Processing and Extractive Metallurgy [17] - part of IMMa
- Mining Technology [18] - part of IMMa
- Natural Materials Association [19] - includes natural fibres
- Packaging Society [20]
- Petroleum & Drilling Engineering [21] - part of IMMa, and includes the Mining Institute of Scotland [22]
- Polymer Society [23] - includes PVC and rubber
- Surface Engineering [24] - corrosion and adhesives
- Sustainable Development Group [25]

History

IOM3 was formed from the merger of the Institute of Materials and the Institution of Mining and Metallurgy in June 2002.

Former organisations

The Institute of Materials itself was a merger of The Institute of Metals, The Institute of Ceramics and The Plastics and Rubber Institute (known as the PRI) in 1993. The latter was itself a merger of the The Plastics Institute and the Institution of the Rubber Industry (known as the IRI) during the 1980s, a reflection of the declining UK rubber manufacturing industry during this period.

Its roots go back to the Iron and Steel Institute which was established in 1869. It received Royal Charters in 1899 and 1975.

More recent mergers include the Institute of Packaging (2005), the Institute of Clay Technology (2006) the Institute of Wood Science (2009) and the Institute of Vitreous Enamellers (2010).

Function

The Institute ensures that courses in materials, minerals and mining technology and engineering conform to the standards for professional registration with the Engineering Council UK, establishes codes of practice and monitors legislative matters affecting members' professional interests.

The professional development programme run by the Institute helps contribute to members' career enhancement towards senior grades of membership and Chartered Engineer (CEng) status.

Members benefit from reduced rates for the Institute's many books, journals and conferences and from access to the Institute's Information Services. These include extensive library resources as well as a team of materials experts who provide consultancy services to Institute members, and to companies who have joined the Institute's Industrial Affiliate Scheme.

Activities

The Institute provides a range of activities and initiatives to benefit the materials and minerals community.

The Institute's educational activities aim to promote the materials discipline to younger generations by allowing access, through the Schools Affiliate Scheme, to a range of educational resources and materials. The Institute has very close links with schools and colleges and is responsible for accrediting university and college courses and industrial training schemes. The Education Department offers teachers courses and teaching resources on materials, as well as careers advice for students. Many Institute publications such as definitive textbooks are available to students at reduced prices. The Institute also offers a series of grants (money) and bursaries to encourage students, and organises events such as the Young Persons' Lecture Competition.

Publications

The Institute's trading subsidiary, IOM Communications Ltd, is responsible for producing the Institute's many books and journals. These include the members' journals, Materials World magazine, The Packaging Professional, Clay Technology, and Wood Focus, and a range of learned journals published by Maney Publishing on behalf of the Institute such as Ironmaking and Steelmaking Journal, Surface Engineering, Powder Metallurgy, Corrosion Engineering, Science and Technology, International Materials Reviews and Materials Science and Technology.

Advice

The Materials Information Service is a service of the Institute which has been giving advice to industry on the selection and use of materials since 1988. This is now part of the Institute's Information Services which includes technical enquiry and library services for the materials, minerals and mining sectors, an information help desk, regionally based advisors and related services. Companies can gain access to the Institute's information resources by joining its Industrial Affiliate Scheme.

Conferences

The Institute's Conference Department organises conferences, events and exhibitions with the Institute's technical committees to help keep members and other delegates informed of the latest developments within the materials, minerals and mining arena. The highlight of the conference calendar is the bi-annual Materials Congress.

Awards

The **Bessemer Gold Medal** is an annual prize awarded by the Institute for "outstanding services to the steel industry". It was established and endowed by Sir Henry Bessemer in 1874. It was first awarded to Isaac Lowthian Bell in 1874. The 2011 award was to Ian Christmas.[26]

References

[1] http://www.iom3.org
[2] The Institute of Materials, Minerals and Mining, Registered Charity no. 269275 (http://www.charitycommission.gov.uk/ShowCharity/RegisterOfCharities/searchresulthandler.aspx?chyno=269275) at the Charity Commission
[3] Photo at Geograph (http://www.geograph.org.uk/photo/1149119)
[4] http://www.iom3.org/content/applied-earth-science-division
[5] http://www.iom3.org/content/automotive-division
[6] http://www.iom3.org/content/biomedical-applications-division
[7] http://www.iom3.org/content/british-composites-society
[8] http://www.iom3.org/content/casting-division
[9] http://www.iom3.org/content/ceramics-society
[10] http://www.iom3.org/content/electronics-division
[11] http://www.iom3.org/content/energy-materials-group
[12] http://www.iom3.org/content/imma-international-mining-minerals-association
[13] http://www.iom3.org/iron-and-steel-society
[14] http://www.iom3.org/content/wood-technology
[15] http://www.iom3.org/content/light-metals-division
[16] http://www.iom3.org/materials-science-technology-division
[17] http://www.iom3.org/content/mineral-processing-extractive-metallurgy-division
[18] http://www.iom3.org/mining-technology-division
[19] http://www.iom3.org/content/natural-materials-association
[20] http://www.iom3.org/content/packaging
[21] http://www.iom3.org/content/petroleum-drilling-engineering-division
[22] http://www.mining-scotland.org
[23] http://www.iom3.org/content/polymer-society
[24] http://www.iom3.org/content/surface-engineering-division
[25] http://www.iom3.org/content/sustainable-development-group
[26] Past Bessemer Gold Medal winners (http://www.iom3.org/content/award-winners-2011)

External links

- Institute of Materials, Minerals and Mining (http://www.iom3.org)

Publications

The Institute produces several magazine titles, Materials World, The Packaging Professional, Wood Focus and Clay Technology. The titles are available to members or by subscription.

- Design exchange: Institute of Materials, Minerals and Mining (http://www.iom3.org/made)
- Design exchange: Materials Knowledge Transfer Network (http://amf.globalwatchonline.com/epicentric_portal/site/AMF/menuitem.06688dadabd92be81006a0108380e1a0/)
- Materials Science: Materials Knowledge Transfer Network (http://www.materialsktn.net)

Video clips

- IOM3 YouTube channel (http://www.youtube.com/user/iom3)

Ostoja_coat_of_arms

Ostoja	
Battle cry: Hostoja, Ostoja	
Details	
Alternative names	Hostoja, Mościc, Ostojczyk
Earliest mention	Moscic 1242, Ostoja 1348
Towns	Konstantynów Łódzki, Terespol, Skierbieszów
Families	**239 families altogether:** Ancyperowicz (Ancyparowicz). Baduski, Balicki, Balinski, Banczelski, Bańkowski (Bankowski), Barankiewicz, Baranowski, Bardecki, Bartkowski (Bartoszewski), Bątkowski, Bębnowski, Bęczelski, Biestrzecki, Blinowski, Błaszkowsk,i Błociszewski, Błyszczanowicz, Bociarski, Bogorajski odm., Bohatyrowicz, Bogusławski, Boguszewski, Boratynski, Bratkowski, Brodowicz, Brokowski, Broniowski, Bukowski, Bzowski. Chełmowski, Chełmski, Chotkowski, Chodorkowski (Unichowski), Chrostecki, Chrostowski, Chrząstowski, Chudziński (Chudzyński), Chyżyński (Chyżewski), Ciechanowski, Cieśliński, Czechowicz, Czeczot, Czernikowski. Danielewicz (Danilewicz), Darowski, Dembowski (Dębowski), Dmosicki (Dmościcki), Dobromirski, Domaracki (Domaradzki), Dubaniewski, Duczman, Duszakiewicz, Dutkowski (Dudkowski), Dworzecki (Dworecki), Dzieczyński. Gajewicz, Gajewski, Gawłowski, Glokman, Głazowski, Głębocki, Głogiński, Głowczewski, Gniady, Godziszewski, Gołębiowski, Gorkowski, Gralewski. Hajewski, Hełczyński, Hrebnicki, Hubnicki, Hulanicki Iłowiecki, Jakliński, Jackowski, Jajkowski, Janiszewski, Jerzykowski, Jotejko (Joteyko), Juckiewicz. Kaczanowski, Kałłaur, Karabczewski (Krabczyjowski, Karapczewski), Karlewicz, Karliński, Kaweczyński (Kawienczyński), Kątkowski (Kontkowski), Kiedrzyński, Kołodziejski, Komorowski (Komor), Kondradzki (Kondracki), Kończycki, Kostkowski, Kotkowski (Chotkowski), Kotnowski, Gołąbek-Kowalski, Kozak, Kozłowiecki, Koźniewski (Kożniewski), Krall, Kreza, Krępski, Krzelczycki (Krzylczycki, Mrożek), Krzelewski, Krzywiec (Okolowicz), Książnicki, Kublicki (Kublicz), Kumanowski, Kurosz. Leski (Lenski, Łęski), Lniski odm.,Łowiecki, Lubochoński (Lubochiński), Lubochowski. Maleczkowski, Marchowski, Marchocki, Markiewicz, Marylski, Miedzwiecki, Mietelski, Miklaszewicz, Miklaszewski, Mikorski, Mikutowicz, Mioduszewski, Modliszewski, Modrzejowski (Modzrzejewski), Mokrzecki, Mościcki. Nagorczewski, Nagórski, Nasierowski, Nieradzki, Nyko (Nyka). Ochocki, Olewiński (Oliwiński), Osieczkowski (Osiczkowski), Ostaniewicz, Ostaszewski, Ostarzewski odm., Ostojski, Owsiany, Ozięblowski.

	Palczyński, Palecki, Pękowski (Pętkowski, Pękosławski), Piaszczyński, Piestrzecki, Pilawski, Podgorski, Podwysocki, Politański, Poniecki, Potapowicz, Potocki, Przeszmiński.
	Raczko, Raczewski, Radogoski (Radogowski, Radogorski), Raduński, Rafalski, Raszewski, Redej (Rhedey), Roguski, Rokosz (Rekosz), Rozkoszewski, Rożniatowski, Rudołtowski, Rudzicki, Rumszewicz, Rylski, Rzepliński, Rzymczykowski.
	Samborski, Samorok (Samorak), Sarbski, Sciborski (Ściborski) , Sędzimir, Siedlecki, Siedlikowski, Siedliski, Siemoński, Sieradzki, Skrzyszewski (Skrzyszowski), Słoński, Słupski, Słuszka, Smuszewski (Smoszewski), Solecki, Stachelski, Stachora (Stachorski, Stachurski), Stackiewicz, Starzewski (Starzeski), Starzycki, Staszewski, Stebelski, Steblecki, Stobiecki, Strzałka (Strzałkowski), Strzelecki, Suchcicki, Suchorabski, Sultzen, Sułecki, Szczoskowski, Swierczyński (Świerczyński), Szabai, Szabajkowicz, Szabay, Szmigłowski, Szyszkowski.
	Targoński, Teliński, Tołkacz, Tomkiewicz, Trzebnicki, Tucznia, Tumiłowicz, Turkułł, Turznicki (Turnicki), Twerbut, Tworowski.
	Ubliński, Unichowski (Chodorkowski), Ulejski (Uleski), Ustarbowski.
	Wadowski, Wasilewski, Wojtkiewicz, Wolski, Woynowski, Woyszyk, Wróblewski Wyrzek, Wysocki (Korzeniewicz).
	Zabierzowski, Zabokrzycki, Zaborowski, Zagórski (Zagorski, Zagurski), Zajarski (Zajerski), Zakrzewski, Zarogowski, Zdaniszewski, Złociszewski (Złoczyszewski), Złoszcz, Żelisławski.

Ostoja - is a Polish coat of arms that originally was used by the family of Ostoja from Kujavia and Greater Poland.[1] As the family expanded their territory to Pomerania, Prussia, Lesser Poland, Slovakia, Hungary, Romania they also adopted several noble families from Lithuania, Belarus and Ukraine, finally turning into a Clan of Ostoja. As different lines of the Clan formed surnames after their properties and adding the adoptions, Ostoja become a Coat of arms used by several families not necessarily blood related with each other.[2] [3] [4]

History

Ostoja coat of arms in Baranów Sandomierski Castle

Blazon

Original version: Coat of arms of the medieval version differed significantly from the generalized form in later times. The following reconstruction appearance comes from Josef Szymanski: *Gules, between an increscent and a decrescent a cross in pale point downwards, all Or. On a helmet a dragon Sable, exhaling fire Gules, on two crescents pointing up, Or. Mantling Sable, lined Or*

Later modified version replaced a cross between crescents with the sword in pale point downwards. On a crowned helmet, five ostrich feathers.

Modified Ostoja in 17th century

Ostoja of Sędzimir

Crown of Count Ostoja

Notable members of the Clan

Notable members[5] [6] [7] [8] of the Clan of Ostoja:

Ladies of Ostoja

- Lady Halina de Krepy (1270) - Daughter of Castellan Piotr Krepy. For the defense of Sandomierz
- Lady Marianna Ścibor-Marchocka (1603–1652) - Author writing prose
- Lady Elżbieta Słuszka (1619–1671) - Lady of great power in Commonwealth
- Lady Krystyna Ścibor-Bogusławska (-1783) - Lady of regality of Wągłczew, nomination received by the King Poniatowski.
- Lady Karolina Wojnarowska (1814–1858) born Rylski - author writing under the pseudonym Karol Nowowiejski
- Lady Emma Ostaszewska (1831–1912) - Social activist, patriot
- Lady Maria A. Ostaszewska (1851–1918) - Social activist, patriot
- Lady Karola Uniechowska(1904–1955) - Voluntary, medical doctor during WWII, participated in the Battle of Monte Cassino
- Lady Zofia Uniechowska (1909–1993) - achieved Virtuti Militari for conspiracy against Nazi government in Poland
- Lady Zofia Rylski - master spy during WWII as "Marie Springer", her information led to localization and destruction of German Battleship Tirpitz
- Lady Izabela Zielińska born Ostaszewska 1910 - musician, decorated with medal of Gloria Artis 2011
- Lady Maja Ostaszewska (born 1972) - theater and film actress. Best Actress Award at the Polish Film Festival held in Gdynia in 1998

Lords and Knights of Ostoja

- Mościc de Magni Kozmin - Voivode of Poznan 1242-1252
- Piotr de Krepy - Castellan of Sandomierz 1270
- Jędrzej of Ostoja - Castellan of Poznan 1343
- Mościc de Stiboricz - Voivode of Gniewkowo 1353
- Jakusz de Blociszewo - Voivode of Lwow 1370
- Abel Biel de Bleszno - Burgrave of Wielun 1376, Lord of regality of Inowrocław and Krzepice. Chamberlain of Wielun
- Stibor de Radzimin - Bishop of Plock 1390
- Mikolaj Bydgoski - Castellan of Bydgoszcz, Hungarian Baron 1400
- Stibor of Stiboricz (1348–1414) - Lord of Slovakia, ruler of Transylvania, Bratislava, Lord on 31 castles, one of richest and most powerful Magnates in Europe
- Moscic de Staszow - Castellan of Poznan 1413
- Swietosław Iłowiecki - Castellan of Karzec 1415
- Andrzej Podczaszy - Voivode of Trencsén, Hungarian Baron
- Stibor de minori Stiboric - Archbishop of Eger
- Rokossius (Rokosz)Chełmski - Bishop of Kamianets-Podilskyi 1378-1398
- Mikolaj Błociszewski - Castellan of Sanok 1403, Lord of Poznan 1417

Crown of Baron Ostoja

Ostoja of Duke Stibor of Stiboricz

Ostoja during early Piast

- Stibor de Beckov (-1434) - Lord of Slovakian Counties, Lord of Orava, Magnate, son of Stibor of Stiboricz, owner of half of western Slovakia, properties in Poland, Morawia and Germany.
- Piotr Chełmski - Burgrave of Krakow 1418, Castellan of Połaniec 1434
- Jan Chełmski - Castellan of Połaniec 1451
- Piotr Gajewski of Ostoja - Castellan of Kalisz 1456
- Piotr Franczoch de Lopuszna - Voivode of Sanok 1456
- Mikolaj Biel de Jerzykowo - Castellan of Ostrow
- Piotr Stibor de Poniec (-1471) - Lord of the regality of Greater Poland 1460 and of Marieburg (Malbork), Diplomat
- Marcin Poniecki (1448–1498) - Son of Piotr Stibor, the first poets writing in polish language
- Jan Jański de Turze - Voivode of Pomerania 1454
- Mikołaj Szarlejski (1400–1457) - Voivode of Inowrocław, Voivode of Kujawy and Brześć Kujawski, Lord of Bydgoszcz, Supreme Commander of polish army in Prussia, Lord of Tuchola County, Brodnica County, Lord of regality in Gniewkowo.
- Piotr de Chotkowo (Kotkowski) - Bishop of Plock 1480-1497
- Marcorius de Magni Solec (Solecki) - Bishop of Sardyensk 1499
- Maciej Kawęczyński (-1572) - Editor, making reforms in Lithuania
- Mikołaj Kreza (-1574) – Rittmeister of the Crown
- Michał Maleczkowski - Magnus procurator of Lesser Poland 1576-1577
- Kacper Karliński (-1590) - Lord of regality of Olsztyn 1563-1587, famouse for defence of Olsztyn in 1587
- Gabriel Słoński (1520–1598) - Architect, Burgrave of Krakowa
- Krzysztof Słuszka - (-1619) Voivode of Livonia (Wendenski), brother of Aleksander, Voivode of Trakai
- Salomon Rysiński (1565–1625) – Famous writer
- Michał Sędziwój (Michael Sendivogius, Sędzimir) (1566–1636) - famous European alchemist, philosopher, medical doctor
- Marcin Szyszkowski (1554–1630) - Bishop of Kraków, Prince of Siewierz
- Wojciech Gajewski - Castellan of Rogozin 1631-1641
- Piotr Szyszkowski - Castellan of Wojno 1643
- Mikołaj Szyszkowski (1590–1643) - Prince-bishop of Warmia from 1633
- Krzysztof Boguszewski (-1635) - painter, artist of Greater Poland
- Stanisław Bzowski – (1567–1637) – member of Dominican Order, friend of reforms, appointed by Vatikan City to write down its history.
- Aleksander Słuszka (1580–1647) - Castellan of Samogitia, Voivode of Minsk (-1638), Novogrod (-1642), Trakai (-1647)
- Paweł Danielewicz, Judge of Vilnius 1648, Marshal of the Lithuanian Court of Justice, Lord of regality of Intursk
- Łukasz Gajewski - Castellan of Santok 1661
- Jan Stachurski - General major 1664
- Kazimierz Siemienowicz (1600–1651) - General of Artilery
- Michał Scibor-Rylski - Castellan of Gostyn 1685

Ostoja in Amorial Gelre

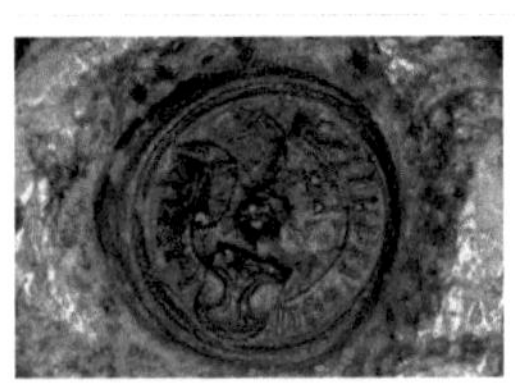
Ostoja, seal from 1466

Beckov castle of Stibor's of Ostoja

- Bogusław Aleksander Unichowski (-1700) - Voivode of Trakai 1696, Marshal of the Lithuanian Tribunal 1687
- Mikołaj Scibor Marchocki - Castellan of Malogoski (Żarnòw) 1697
- Josef Bogusław Sluszka (1652–1701) - Hetman, Castellan of Trakai and Vilnius
- Dominik Michał Słuszka (1655–1713) - Voivode of Polotsk
- Aleksander Jozef Unichowski (-1725) - Castellan of Samogitia (Zmudz)
- Aleksander Scibor Marchocki (-1737) - Castellan of Malogoski
- Franciszek Gajewski (Castellan)) (1675–1753) - Castellan of konarsk-kuyavia
- Florian Hrebnicki (1684-1762) - Archbishop of Polotsk
- Antoni Gajewski (-1775) - Castellan of Naklo, Lord of the regality of Łęczyca and of Kościany
- Rafał Tadeusz Gajewski (1714–1776)- Castllan of Rogozin
- Józef Jakliński - Castellan of Kamensk/Spicymir 1759-1775
- Józef Siemoński - General adj. of King Poniatowski, supreeme commander of Sandomierz uprising iniciated by Kościuszko
- Karol Podgorski (-1781)- Major General in Prussian army
- Michał Władysław Lniski (1723–1777) - vice Voivode and Marshal of the Contederetion of Bar in Pomerania
- Wojciech Marchocki (-1788) - Castellan of Sanok
- Józef Andrzej Mikorski - Castellan in Rawy from 1791
- Franciszek Ksawery Ścibor-Bogusławski (1713-1796) - Rittmeister of the Confederation of Bar
- Tadeusz Błociszewski (-1803) – Major General
- Michał Ostaszewski (1720-1816) - one of main initiators of the Confederation of Bar in Subcarpathian Voivodeship
- Tomasz Ostaszewski(1746-1817) - Bishop of Plock
- Antoni Baranowski (general) (1760–1821) – Major General
- Józef Siemoński – general adj. of King Poniatowski
- Ignacy Ścibor Marchocki (1755–1827) - "Kingdom of Mińkowce"
- Jan Czeczot (1796–1847) - Poet, ethnographer
- Andrzej Miklaszewski - Actual State Councillor (eg. Marshall and General - Table of Ranks) of Russia 1863
- Jan Kazimierz Ordyniec (1797–1863) - great patriot, publisher of "Dziennik Warszawski"
- Spirydion Ostaszewski (1797–1875) - for writing down polish legends
- Kacper Kotkowski (1814–1875) - catholic priest, head and commissar of the Sandomierz uprising

- Stanisław Błociszewski (1804–1888) - Officer, great patriot, Virtuti Militari
- Teofil Wojciech Ostaszewski (1807-1889) - initiated first program against Serfdom, Marshal of Brzostowo County
- Łukasz Solecki (1827–1900) - Bishop of Przemyśl, Professor of the Lviv University
- Jan Aleksander Karłowicz (1836–1903) - Ethnographer, linguist, folklore
- Mieczysław Karłowicz (1876–1909) – Composer of several symphonies and poems
- Zygmunt Czechowicz (1831–1907) - One of the initiators of the uprising of the Belarus Nation
- Hipotit Brodowicz - Major General WCR
- Stanisław Ostaszewski (1862–1915) – Inventor, promotor of new technology
- Adam Mokrzecki (1856–1921) - Major General in Russian army, General in polish army
- Rudolf Starzewski – Famous chief editor of "Czasu", described by Wyspiański in "Wedding"
- Adam Mokrzecki (1856–1921) - Major General in russian and later in polish army
- Stefan Mokrzecki (1862–1932) - General in russian and later inpolish army
- Witold Ścibor-Rylski (1871–1926) - General
- Władysław Chotkowski (1843–1926) – Professor and head of Jagiellonian University
- Włodzimierz Zagórski (1882–1927) - General
- Adam Ostaszewski (1860–1934) - Inventor, Pioneer
- Adam Ostaszewski (1866-1934) - President of Plock
- Bronisław Bohatyrewicz (1870–1940) - General
- Adam Hrebnicki-Doktorowicz (1857–1941) – Professor in agriculture development, founder if Institute.
- Karzimierz Zagórski (1883–1944) - Adventurer-Pioneer
- Zygmund Ignacy Rylski (1898-1945) - legendary *Major Hańcza*, later advanced to rang of Colonel.
- Stanisław Chrostowski (1897–1947) - Professor, artist
- Wacław Krzywiec (1908–1956) – famous komandor of the legendary ORP Błyskawica warship
- Maxim Rylski, (1895-1969) - famous poet in Ukraine
- Stanisław Doktorowicz-Hrebnicki (1888–1974) - Professor in geology, decorated.
- Antoni Uniechowski (1903–1976)- Famous painter, drawings
- Stefan Ścibor-Bogusławski (1897-1978) - Colonel, richly awarded, also for contributing in the Battle of Monte Cassino
- Bronisław Hełczyński (1890–1978) - Professor in law, Oxford University, minister
- Aleksander Ścibor-Rylski (1928-1983) - poet, writer and film director
- Tadeusz Sędzimir (1894–1989) - World wide known inventor
- Józef Stanisław Ostoja-Kotkowski (1922–1994) - Artist
- Tadeusz Ostaszewski (1918–2003) - Professor of fine arts, Krakow University
- Adam Kozłowiecki (1911–2007) - Archbishop of the Archdiocese of Lusaka in Zambia.
- Andrzej Zagórski (1926–2007) - devoted officer in Armia Krajowa, wrote over 250 publications about polish underground resistance
- Kazimierz Tumiłowicz (1932–2008) – great patriot, creator of Siberian association of remembrance, social worker in Greater Poland
- Andrzej Ostoja-Owsiany (1931-2008) - Senator of Poland after the fall of communism
- Zbigniew Ścibor Rylski (born 1917) - General
- Zbigniew Rylski (born 1923) - widely decorated patriot, Major in polish army
- Eustachy Rylski (born 1944) - widely recognized drama and prose writer

Variants

- Variants[9] : Bogorajski, Gawłowski, Kleczewski, Miklaszewski, Mokrzewski, Korzeniewicz-Wysocki, Lniski, Marylski, Nagórski, Okołowicz Orda, Osiecki, Ostarzewski, Plat, Pokroszyński, Racięski, Sędzimir, Strzalkowski, Turkul and Wasilewski. The Przegonia coat of arms origin from Ostoja.

See also

- Polish heraldry
- Heraldry
- Coat of arms
- Clan of Ostoja

References

[1] Andrzej Kulikowski: Wielki Herbarz Rodów Polskich. Warszawa: Świat Książki-Bertelsmann Media, 2005. ISBN 83-7391-523-0
[2] Seweryn Uruski: Rodzina.Herbarz szlachty polskiej, Warszawa 1906
[3] Franciszek Piekosiński: Heraldyka polska wieków średnich. Kraków, Akademia Umiejętności 1899
[4] Adam Boniecki: Herbarz polski. T. 16. Warszawa: skł. gł. Gebethner i Wolff, 1913
[5] Adam Boniecki: Herbarz polski. T. 16. Warszawa: skł. gł. Gebethner i Wolff, 1913
[6] Seweryn Uruski: Rodzina .Herbarz szlachty polskiej, Warszawa 1906
[7] Polska Akademia Nauk, "Polski Slownik Biograficzny" (Polish Biographical Dictionary), Krakow from 1935
[8] Adam Boniecki: Herbarz polski. T. 16. Warszawa: skł. gł. Gebethner i Wolff, 1913
[9] Tadeusz Gajl: Herbarz polski od średniowiecza do XX wieku,L&L, 2007, ISBN 978-83-60597-10-1

External links

- Adam Boniecki, Herbarz (http://ebuw.uw.edu.pl/dlibra/doccontent?id=136&dirids=1)
- type Uruski - Seweryn Uruski, Herbarz szlachty polskiej (http://www.wbc.poznan.pl/dlibra)
- Tadeusz Gajl, Herbarz (http://www.herbyszlachty.pl/)
- Ostoja Clan Association on Facebook (http://www.facebook.com/group.php?gid=192794509605)

Sources

- Adam Boniecki: Herbarz polski. T. 16. Warszawa: skł. gł. Gebethner i Wolff, 1913
- Alfred Znamierowski: Herbarz rodowy. Warszawa: Świat Książki, ISBN 83-7391-166-9
- Tadeusz Gajl: Herbarz polski od średniowiecza do XX wieku,L&L, 2007, ISBN 978-83-60597-10-1
- Seweryn Uruski: Rodzina.Herbarz szlachty polskiej, Warszawa 1906
- Kasper Niesiecki: Herbarz polski, 1839–1846
- Jan Długosz: Jana Długosza kanonika krakowskiego Dziejów polskich ksiąg dwanaście, Kraków 1867-1870
- Józef Szymański: Herbarz rycerstwa polskiego z XVI wieku. Warszawa, ISBN 83-7181-217-5
- Andrzej Kulikowski: Wielki Herbarz Rodów Polskich. Warszawa: Świat Książki-Bertelsmann Media, 2005. ISBN 83-7391-523-0
- Medieval armorial: Gelre Armorial
- Medieval armorials: Armorial "Zlotego Runa", Armorial Lyncenich, Codex Bergshammar, "Klejnoty" of Jan Długosz, Armorial Bellenville
- Franciszek Piekosiński: Heraldyka polska wieków średnich. Kraków, Akademia Umiejętności 1899

Cross_of_Merit_(Poland)

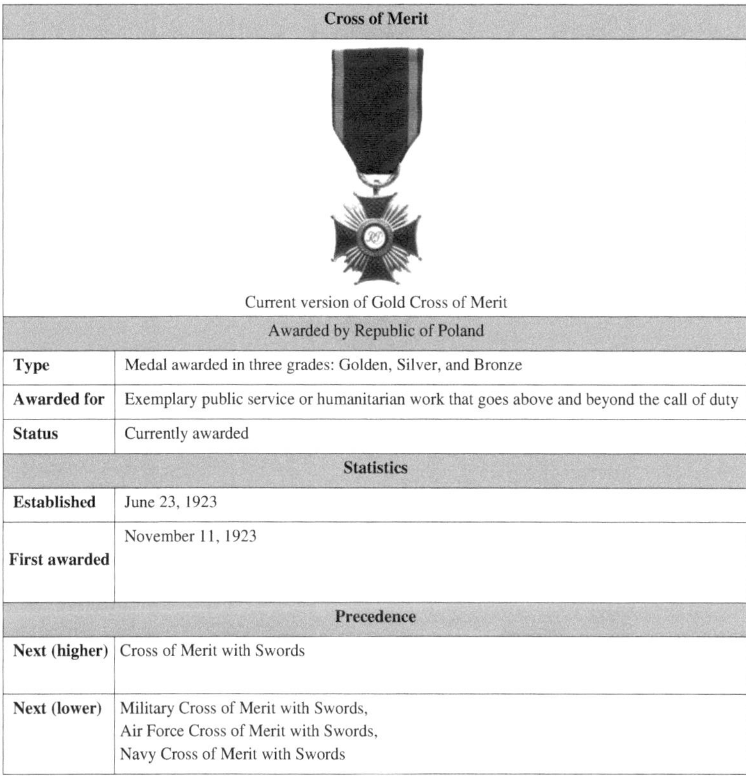

Cross of Merit	
Current version of Gold Cross of Merit	
Awarded by Republic of Poland	
Type	Medal awarded in three grades: Golden, Silver, and Bronze
Awarded for	Exemplary public service or humanitarian work that goes above and beyond the call of duty
Status	Currently awarded
Statistics	
Established	June 23, 1923
First awarded	November 11, 1923
Precedence	
Next (higher)	Cross of Merit with Swords
Next (lower)	Military Cross of Merit with Swords, Air Force Cross of Merit with Swords, Navy Cross of Merit with Swords

The **Cross of Merit** (Polish: *Krzyż Zasługi*) is a Polish civil state award established on June 23, 1923, to recognize services to the state.

The Order has three grades:

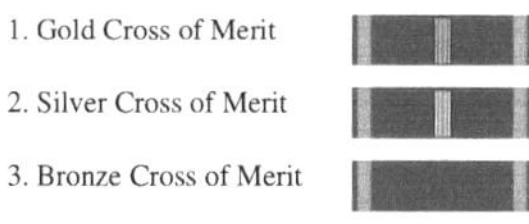

1. Gold Cross of Merit
2. Silver Cross of Merit
3. Bronze Cross of Merit

At the time of its establishment in 1923, the Cross of Merit was the highest civilian award in Poland. It was awarded to citizens who went beyond the call of duty in their work for the country and society as a whole.

See also

- Cross of Merit (Polish Scouting and Guiding Association)

Royal_Swedish_Academy_of_Sciences

Royal Swedish Academy of Sciences	
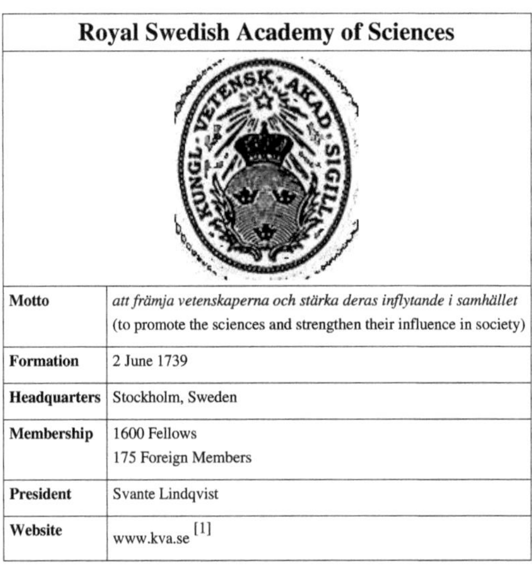	
Motto	*att främja vetenskaperna och stärka deras inflytande i samhället* (to promote the sciences and strengthen their influence in society)
Formation	2 June 1739
Headquarters	Stockholm, Sweden
Membership	1600 Fellows 175 Foreign Members
President	Svante Lindqvist
Website	www.kva.se [1]

Kungliga Vetenskapsakademien

The **Royal Swedish Academy of Sciences** or *Kungliga Vetenskapsakademien* ("KVA") is one of the Royal Academies of Sweden. The Academy is an independent, non-governmental scientific organization which acts to promote the sciences, primarily the natural sciences and mathematics.

The Academy was founded on 2 June 1739 by naturalist Carl Linnaeus, mercantilist Jonas Alströmer, mechanical engineer Mårten Triewald, civil servants Sten Carl Bielke and Carl Wilhelm Cederhielm, and politician Anders Johan von Höpken.[2]

The purpose of the academy was to focus on practically useful knowledge, and to publish in Swedish in order to widely disseminate the academy's findings. The academy was intended to be different from the Royal Society of Sciences in Uppsala, which had been founded in 1719 and published in Latin. The location close to the commercial activities in Sweden's capital (which unlike Uppsala did not have a university at this time) was also intentional. The academy was modeled after the Royal Society of London and Academie Royale des Sciences in Paris, France, which some of the founding members were familiar with.

Committees of the Academy act as selection boards for international prizes:

- Nobel Prizes in Physics and Chemistry[3]
- Sveriges Riksbank Prize in Economic Sciences in Memory of Alfred Nobel[4] (also known as the Nobel Memorial Prize in Economics)
- Crafoord Prizes in astronomy and mathematics, geosciences, biosciences (with an emphasis on ecology), and polyarthritis (rheumatoid arthritis)[5]

- Rolf Schock Prizes in logic and philosophy[6]
- Gregori Aminoff Prize in crystallography[7]
- Oskar Klein medal

and national prizes:[8]

- Göran Gustafsson Prizes for research in the natural sciences and medicine
- Söderberg Prize in economics or jurisprudence
- Tage Erlander Prize in physics, chemistry, technology, and biology
- Ingvar Lindqvist Prizes for teachers in the fields of physics, chemistry, biology and mathematics.

List of permanent secretaries

The following persons have served as permanent secretaries of the Academy:

- Anders Johan von Höpken, 1739–1740, 1740–1741
- Augustin Ehrensvärd, April–June 1740
- Jacob Faggot, 1741–1744
- Pehr Elvius, 1744–1749
- Pehr Wilhelm Wargentin, 1749–1783
- Johan Carl Wilcke and Henrik Nicander, 1784–1796
- Daniel Melanderhjelm and Henrik Nicander, 1796–1803
- Jöns Svanberg and Carl Gustaf Sjöstén 1803-1808; Sjöstén was removed 1808 for negligence of his duties
- Jöns Svanberg, 1809–1811
- Olof Swartz, 1811–1818
- Jöns Jacob Berzelius, 1818–1848
- Peter Fredrik Wahlberg, 1848–1866
- Georg Lindhagen, 1866–1901
- Christopher Aurivillius, 1901–1923
- Henrik Gustaf Söderbaum, 1923–1933
- Henning Pleijel, 1933–1943
- Arne Westgren, 1943–1959
- Erik Rudberg, 1959–1972
- Carl Gustaf Bernhard, 1973–1981
- Tord Ganelius, 1981–1989
- Carl-Olof Jacobson, 1989–1997
- Erling Norrby, 1997 - 30 June 2003
- Gunnar Öquist, 1 July 2003 - 30 June 2010
- Staffan Normark, appointed from 1 July 2010

Publications

The transactions of the Academy (*Vetenskapsakademiens handlingar*) were published as its main series between 1739 and 1974. In parallel, other major series have appeared and gone:

- *Öfversigt af Kungl. Vetenskapsakademiens förhandlingar* (1844–1903)
- *Bihang till Vetenskapsakademiens Handlingar* (1872–1902)
- *Vetenskapsakademiens årsbok* (1903–1969)

The Academy started to publish annual reports in physics and chemistry (1826), technology (1827), botany (1831), and zoology (1832). These lasted into the 1860s, when they were replaced by the single *Bihang* series (meaning: supplement to the transactions). Starting in 1887, this series was once again split into four sections (*afdelning*), which in 1903 became independent scientific journals of their own, titled "Arkiv för..." (archive for...), among them

- *Arkiv för matematik, astronomi och fysik* (1903–1949).

Kongl. Svenska Vetenskaps-Academiens handlingar, volume XI (1750).

Further restructuring of their topics occurred in 1949 and 1974.

The Academy's first online-only (born digital) journal is *Electronic Transactions on Artificial Intelligence* or "ETAI" (ISSN 1403-3534). It was founded in 1997 by Erik Sandewall, professor of computer science at Linköping University.

Current publications

- *Ambio* (1972-)
- *Acta Mathematica* (1882-)
- *Arkiv för matematik* (1949- with this title; 1903-1949 also including physics and astronomy)
- *Acta Zoologica* (1920-)
- *Levnadsteckningar över Vetenskapsakademiens ledamöter* (1869-), biographies of deceased members
- *Physica Scripta* (1970-), jointly with the Norwegian Academy of Science and Letters
- *Porträttmatrikel* (1971-), portraits of current members
- *Zoologica Scripta* (1972-), jointly with the Norwegian Academy of Science and Letters
- *ETAI* (1997-)

See also

- Members of the Royal Swedish Academy of Sciences

The Royal Swedish Academy of Sciences

References

[1] http://www.kva.se/en/
[2] "History" (http://www.kva.se/en/About-the-academy/History/). The Royal Swedish Academy of Sciences. . Retrieved 2009-10-18.
[3] "Nobel Prizes" (http://www.kva.se/en/Prizes/Nobelprises/). The Royal Swedish Academy of Sciences. . Retrieved 2009-10-18.
[4] "Prize in Economic Sciences" (http://www.kva.se/en/Prizes/Prize-in-Economic-Sciences/). The Royal Swedish Academy of Sciences. . Retrieved 2009-10-18.
[5] "Crafoord Prize" (http://www.kva.se/en/Prizes/Crafoord-Prize/). The Royal Swedish Academy of Sciences. . Retrieved 2009-10-18.

[6] "Rolf Schock Prizes" (http://www.kva.se/en/Prizes/Rolf-Shock-prisen/). The Royal Swedish Academy of Sciences. . Retrieved 2009-10-18.

[7] "Gregori Aminoff Prize" (http://www.kva.se/en/Prizes/Gregori-Aminoff-Prize/). The Royal Swedish Academy of Sciences. . Retrieved 2009-10-18.

[8] "National prizes" (http://www.kva.se/KVA_Root/eng/awards/national/index.asp?br=ns&ver=6up). The Royal Swedish Academy of Sciences. . Retrieved 2007-10-18.

External links

- The Royal Swedish Academy of Sciences (http://www.kva.se/en/)
- ETAI (http://www.etaij.org/), the journal's website.

Statue_of_Liberty

Statue of Liberty	
Location:	Liberty Island New York City, New York, U.S.[1]
Coordinates:	40°41′21″N 74°2′40″W
Height:	151 feet 1 inch (46 meters) Ground to torch: 305 feet 1 inch (93 meters)
Dedicated:	October 28, 1886
Restored:	1938, 1984–1986, 2011–2012 (ongoing)
Sculptor:	Frédéric Auguste Bartholdi
Visitation:	3.2 million (in 2009[2])
Governing body:	U.S. National Park Service
UNESCO World Heritage Site	
Type:	Cultural
Criteria:	i, vi
Designated:	1984 (8th session)
Reference #:	307 [3]
State Party:	United States
Region:	Europe and North America
U.S. National Register of Historic Places	
Official name: Statue of Liberty National Monument, Ellis Island and Liberty Island	
Designated:	October 15, 1966[4]
Reference #:	66000058
U.S. National Monument	
Designated:	October 15, 1924
Designated by:	President Calvin Coolidge[5]
New York City Landmark	
Type:	Individual
Designated:	September 14, 1976[6]

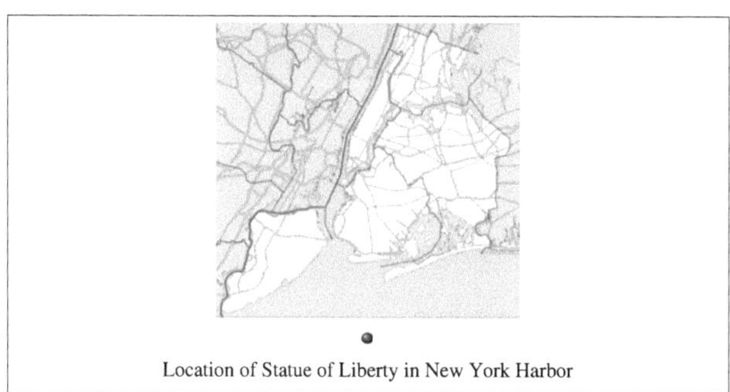

Location of Statue of Liberty in New York Harbor

The **Statue of Liberty** (***Liberty Enlightening the World***; French: *La Liberté éclairant le monde*) is a colossal neoclassical sculpture on Liberty Island in New York Harbor, designed by Frédéric Bartholdi and dedicated on October 28, 1886. The statue, a gift to the United States from the people of France, is of a robed female figure representing Libertas, the Roman goddess of freedom, who bears a torch and a *tabula ansata* (a tablet evoking the law) upon which is inscribed the date of the American Declaration of Independence, July 4, 1776. A broken chain lies at her feet. The statue has become an icon of freedom and of the United States.

Bartholdi was inspired by French law professor, politician, and anti-slavery activist Édouard René de Laboulaye, who commented in 1865 that any monument raised to American independence would properly be a joint project of the French and American peoples. Due to the troubled political situation in France, work on the statue did not commence until the early 1870s. In 1875, Laboulaye proposed that the French finance the statue and the Americans provide the pedestal and the site. Bartholdi completed the head and the torch-bearing arm before the statue was fully designed, and these pieces were exhibited for publicity at international expositions. The arm was displayed at the Centennial Exposition in 1876 and in New York's Madison Square Park from 1876 to 1882. Fundraising proved difficult, especially for the Americans, and by 1885 work on the pedestal was threatened due to lack of funds. Publisher Joseph Pulitzer of the *World* started a drive for donations to complete the project that attracted more than 120,000 contributors, most of whom gave less than a dollar. The statue was constructed in France, shipped overseas in crates, and assembled on the completed pedestal on what was then called Bedloe's Island. The statue's completion was marked by New York's first ticker-tape parade and a dedication ceremony presided over by President Grover Cleveland.

The statue was administered by the United States Lighthouse Board until 1901 and then by the Department of War; since 1933 it has been maintained by the National Park Service. The statue was closed for renovation for much of 1938. In the early 1980s, it was found to have deteriorated to such an extent that a major restoration was required. While the statue was closed from 1984 to 1986, the torch and a large part of the internal structure were replaced. After the September 11 attacks in 2001, it was closed for reasons of safety and security; the pedestal reopened in 2004 and the statue in 2009, with limits on the number of visitors allowed to ascend to the crown. The statue, including the pedestal and base, closed beginning on October 29, 2011 for up to a year, so that a secondary staircase and other safety features can be installed; Liberty Island remains open. Public access to the balcony surrounding the torch has been barred for safety reasons since 1916.

Design and construction

Origin

The origin of the Statue of Liberty project is generally traced to a comment made by French law professor and politician Édouard René de Laboulaye in mid-1865. In after-dinner conversation at his home near Versailles, Laboulaye, an ardent supporter of the Union in the American Civil War, stated: "If a monument should rise in the United States, as a memorial to their independence, I should think it only natural if it were built by united effort—a common work of both our nations."[7]

Bartholdi's design patent

Laboulaye's comment was not intended as a proposal, but it inspired a young sculptor, Frédéric Bartholdi, who was present at the dinner.[7] Given the repressive nature of the regime of Napoleon III, Bartholdi took no immediate action on the idea except to discuss it with Laboulaye. Instead, Bartholdi approached Ismail Pasha, Khedive of Egypt, with a plan to build a huge lighthouse in the form of an ancient Egyptian female *fellah* or peasant, robed and holding a torch aloft, at the northern entrance to the Suez Canal in Port Said. Sketches and models were made of the proposed work, though it was never erected. There was a classical precedent for the Suez proposal, the Colossus of Rhodes: an ancient bronze statue of the Greek god of the sun, Helios. This statue is believed to have been over 100 feet (30 m) high, and it similarly stood at a harbor entrance and carried a light to guide ships.[8]

The American project was further delayed by the Franco-Prussian War, in which Bartholdi served as a major of militia.[7] In the war, Napoleon III was captured and deposed. Bartholdi's home province of Alsace was lost to the Prussians, and a more liberal republic was installed in France.[7] As Bartholdi had been planning a trip to the United States, he and Laboulaye decided the time was right to discuss the idea with influential Americans.[9] In June 1871, Bartholdi crossed the Atlantic, with letters of introduction signed by Laboulaye.[10]

Arriving at New York Harbor, Bartholdi focused on Bedloe's Island as a site for the statue, struck by the fact that vessels arriving in New York had to sail past it. He was delighted to learn that the island was owned by the United States government—it had been ceded by the New York State Legislature in 1800 for harbor defense. It was thus, as he put it in a letter to Laboulaye: "land common to all the states."[11] As well as meeting many influential New Yorkers, Bartholdi visited President Ulysses S. Grant, who assured him that it would not be difficult to obtain the site for the statue.[12] Bartholdi crossed the United States twice by rail, and met many Americans whom he felt would be sympathetic to the project.[10] But he remained concerned that popular opinion on both sides of the Atlantic was insufficiently supportive of the proposal, and he and Laboulaye decided to wait before mounting a public campaign.[13]

Bartholdi's *Lion of Belfort*

Bartholdi had made a first model of his concept in 1870.[14] The son of a friend of Bartholdi's, American artist John La Farge, later maintained that Bartholdi made the first sketches for the statue during his U.S. visit at La Farge's Rhode Island studio. Bartholdi continued to develop the concept following his return to France.[14] He also worked on a number of sculptures designed to bolster French patriotism after the defeat by the Prussians. One of these was the *Lion of Belfort*, a monumental sculpture carved in sandstone below the fortress of Belfort, which during the war had resisted a Prussian siege for over three months. The defiant lion, 73 feet (22 m) long and half that in height, displays an emotional quality characteristic of Romanticism, which Bartholdi would later bring to the Statue of Liberty.[15]

Design, style, and symbolism

Detail from a fresco by Constantino Brumidi in the U.S. Capitol in Washington, D.C., showing two early symbols of America: Columbia (left) and the Indian princess

Bartholdi and Laboulaye considered how best to express the idea of American liberty.[16] In early American history, two female figures were frequently used as cultural symbols of the nation.[17] One, Columbia, was seen as an embodiment of the United States in the manner that Britannia was identified with the United Kingdom and Marianne came to represent France. Columbia had supplanted the earlier figure of an Indian princess, which had come to be regarded as uncivilized and derogatory toward Americans.[17] The other significant female icon in American culture was a representation of Liberty, derived from Libertas, the goddess of freedom widely worshipped in ancient Rome, especially among emancipated slaves. A Liberty figure adorned most American coins of the time,[16] and representations of Liberty appeared in popular and civic art, including Thomas Crawford's *Statue of Freedom* (1863) atop the dome of the United States Capitol Building.[16]

Artists of the 18th and 19th centuries striving to evoke republican ideals commonly used representations of Liberty.[16] A figure of Liberty was also depicted on the Great Seal of France.[16] However, Bartholdi and Laboulaye avoided an image of revolutionary liberty such as that depicted in Eugène Delacroix's famed *Liberty Leading the People* (1830). In this painting, which commemorates France's Revolution of 1830, a half-clothed Liberty leads an armed mob over the bodies of the fallen.[17] Laboulaye had no sympathy for revolution, and so Bartholdi's figure would be fully dressed in flowing robes.[17] Instead of the impression of violence in the Delacroix work, Bartholdi wished to give the statue a peaceful appearance and chose a torch, representing progress, for the figure to hold.[4]

Crawford's statue was designed in the early 1850s. It was originally to be crowned with a *pileus*, the cap given to emancipated slaves in ancient Rome. Secretary of War Jefferson Davis, a Southerner who would later serve as president of the Confederate States of America, was concerned that the *pileus* would be taken as an abolitionist symbol. He ordered that it be changed to a helmet.[18] Delacroix's figure wears a *pileus*,[17] and Bartholdi at first considered placing one on his figure as well. Instead, he used a diadem, or crown, to top its head.[19] In so doing, he avoided a reference to Marianne, who invariably wears a *pileus*.[20] The seven rays form a halo or aureole.[21] They evoke the sun, the seven seas, and the seven continents,[22] and represent another means, besides the torch, whereby Liberty enlightens the world.[4]

Bartholdi's early models were all similar in concept: a female figure in neoclassical style representing liberty, wearing a *stola* and *pella* (gown and cloak, common in depictions of Roman goddesses) and holding a torch aloft. The face was modeled after that of Charlotte Beysser Bartholdi, the sculptor's mother.[23] He designed the figure with a strong, uncomplicated silhouette, which would be set off well by its dramatic harbor placement and allow passengers on vessels entering New York Bay to experience a changing perspective on the statue as they proceeded toward Manhattan. He gave it bold classical contours and applied simplified modeling, reflecting the huge scale of the project and its solemn purpose.[4] Bartholdi wrote of his technique:

> The surfaces should be broad and simple, defined by a bold and clear design, accentuated in the important places. The enlargement of the details or their multiplicity is to be feared. By exaggerating the forms, in order to render them more clearly visible, or by enriching them with details, we would destroy the proportion of the work. Finally, the model, like the design, should have a summarized character, such as one would give to a rapid sketch. Only it is necessary that this character should be the product of volition and study, and that the artist, concentrating his knowledge, should find the form and the line in its greatest simplicity.[24]

Thomas Crawford's *Statue of Freedom*

Bartholdi made alterations in the design as the project evolved. Bartholdi considered having Liberty hold a broken chain, but decided this would be too divisive in the days after the Civil War. The erected statue does rise over a broken chain, half-hidden by her robes and difficult to see from the ground.[19] Bartholdi was initially uncertain of what to place in Liberty's left hand; he settled on a *tabula ansata*, a keystone-shaped tablet[25] used to evoke the concept of law.[26] Though Bartholdi greatly admired the United States Constitution, he chose to inscribe "JULY IV MDCCLXXVI" on the tablet, thus associating the date of the country's Declaration of Independence with the concept of liberty.[25]

Consultations with the metalwork foundry Gaget, Gauthier & Co. led Bartholdi to conclude that the skin should be made of copper sheets, beaten to shape by the repoussé method.[27] An advantage of this choice was that the entire statue would be light for its volume—the copper need be only .094 inches (2.4 mm) thick. He decided on a height of 151 feet (46 m) for the statue, double that of Italy's Colosso di San Carlo Borromeo and the German statue of Arminius, both made with the same method.[28] Bartholdi interested a former teacher of his, architect Eugène Viollet-le-Duc, in the project. Viollet-le-Duc planned to construct a brick pier within the statue, to which the skin would be anchored.[29]

Announcement and early work

In 1875, France was enjoying improved political stability and a recovering postwar economy. Growing interest in the upcoming Centennial Exhibition in Philadelphia led Laboulaye to decide it was time to seek public support.[30] In September 1875, he announced the project and the formation of the Franco-American Union as its fundraising arm. With the announcement, the statue was given a name, *Liberty Enlightening the World*.[31] The French would finance the statue; Americans would be expected to pay for the pedestal.[32] The announcement provoked a generally favorable reaction in France, though many Frenchmen resented the United States for not coming to their aid during the war with Prussia.[31] French monarchists opposed the statue, if for no other reason than it was proposed by the liberal Laboulaye, who had recently been elected a senator for life.[32] Laboulaye arranged events designed to appeal to the rich and powerful, including a special performance at the Paris Opera on April 25, 1876, that featured a new cantata by composer Charles Gounod. The piece was titled *La Liberté éclairant le monde*, the French version of the statue's announced name.[31]

Stereoscopic image of right arm and torch of the Statue of Liberty, 1876 Centennial Exposition

Despite its initial focus on the elites, the Union was successful in raising funds from across French society. Schoolchildren and ordinary citizens gave, as did 181 French municipalities. Laboulaye's political allies supported the call, as did descendants of the French contingent in the American Revolutionary War. Less idealistically, contributions came from those who hoped for American support in the French attempt to build the Panama Canal. The firm of Japy Frères, copper merchants, donated all the copper needed to build the statue, a gift valued at 64,000 francs (about $16,000 at the time or the equivalent of $323,000 today).[33] [34] The copper is said to have come from a mine in Visnes, Norway,[35] though this has not been conclusively determined.[36]

Although plans for the statue had not been finalized, Bartholdi moved forward with fabrication of the right arm, bearing the torch, and the head. Work began at the Gaget, Gauthier & Co. workshop.[37] In May 1876, Bartholdi traveled to the United States as a member of a French delegation to the Centennial Exhibition,[38] and arranged for a huge painting of the statue to be shown in New York as part of the Centennial festivities.[39] The arm did not arrive in Philadelphia until August; because of its late arrival, it was not listed in the exhibition catalogue, and while some reports correctly identified the work, others called it the "Colossal Arm" or "Bartholdi Electric Light". The exhibition grounds contained a number of monumental artworks to compete for fairgoers' interest, including an outsized fountain designed by Bartholdi.[40] Nevertheless, the arm proved popular in the exhibition's waning days, and visitors would climb up to the balcony of the torch to view the fairgrounds.[41] After the exhibition closed, the arm was transported to New York, where it remained on display in Madison Square Park for several years before it was returned to France to join the rest of the statue.[41]

During his second trip to the United States, Bartholdi addressed a number of groups about the project, and urged the formation of American committees of the Franco-American Union.[42] Committees to raise money to pay for the foundation and pedestal were formed in New York, Boston, and Philadelphia.[43] The New York group eventually took on most of the responsibility for American fundraising and is often referred to as the "American Committee".[44] One of its members was 19-year-old Theodore Roosevelt, the future governor of New York and president of the United States.[42] On March 3, 1877, on his final full day in office, President Grant signed a joint resolution that authorized the President to accept the statue when it was presented by France and to select a site for it. President Rutherford B. Hayes, who took office the following day, selected the Bedloe's Island site that Bartholdi had proposed.[45]

Construction in France

The statue's head on exhibit at the Paris World's Fair, 1878

On his return to Paris in 1877, Bartholdi concentrated on completing the head, which was exhibited at the 1878 Paris World's Fair. Fundraising continued, with models of the statue put on sale. Tickets to view the construction activity at the Gaget, Gauthier & Co. workshop were also offered.[34] The French government authorized a lottery; among the prizes were valuable silver plate and a terracotta model of the statue. By the end of 1879, about 250,000 francs had been raised.[46]

The head and arm had been built with assistance from Viollet-le-Duc, who fell ill in 1879. He soon died, leaving no indication of how he intended to transition from the copper skin to his proposed masonry pier.[47] The following year, Bartholdi was able to obtain the services of the innovative designer and builder Alexandre-Gustave Eiffel.[34] Eiffel and his structural engineer, Maurice Koechlin, decided to abandon the pier and instead build an iron truss tower. Eiffel opted not to use a completely rigid structure, which would force stresses to accumulate in the skin and lead eventually to cracking. To enable the statue to move slightly in the winds of New York Harbor and as the metal expanded on hot summer days, he loosely connected the support structure to the skin using an armature—a metal framework that ends in a mesh of metal straps, known as "saddles", that are riveted to the skin, providing firm support. In a labor-intensive process, each saddle had to be crafted individually.[48] To prevent galvanic corrosion between the copper skin and the iron support structure, Eiffel insulated the skin with asbestos impregnated with shellac.[49] The change in structural material from masonry to iron allowed Bartholdi to change his plans for the statue's assembly. He had originally expected to assemble the skin on-site as the masonry pier was built; instead he decided to build the statue in France and have it disassembled and transported to the United States for reassembly in place on Bedloe's Island.[50]

Eiffel's design made the statue one of the earliest examples of curtain wall construction, in which the exterior of the structure is not load bearing, but is instead supported by an interior framework. He included two interior spiral staircases, to make it easier for visitors to reach the observation point in the crown.[51] Access to an observation platform surrounding the torch was also provided, but the narrowness of the arm allowed for only a single ladder, 40 feet (12 m) long.[52] As the pylon tower arose, Eiffel and Bartholdi coordinated their work carefully so that completed segments of skin would fit exactly on the support structure.[53]

In a symbolic act, the first rivet placed into the skin, fixing a copper plate onto the statue's big toe, was driven by United States Ambassador to France Levi P. Morton.[54] The skin was not, however, crafted in exact sequence from low to high; work proceeded on a number of segments simultaneously in a manner often confusing to visitors.[55] Some work was performed by contractors—one of the fingers was made to Bartholdi's exacting specifications by a coppersmith in the southern French town of Montauban.[56] By 1882, the statue was complete up to the waist, an event Barthodi celebrated by inviting reporters to lunch on a platform built within the statue.[57] Laboulaye died in 1883. He was succeeded as chairman of the French committee by Ferdinand de Lesseps, builder of the Suez Canal. The completed statue was formally presented to Ambassador Morton at a ceremony in Paris on July 4, 1884, and de Lesseps announced that the French government had agreed to pay for its transport to New York.[58] The statue remained intact in Paris pending sufficient progress on the pedestal; by January 1885, this had occurred and the statue was disassembled and crated for its ocean voyage.[59]

Fundraising, criticism, and construction in the United States

The committees in the United States faced great difficulties in obtaining funds. The Panic of 1873 had led to an economic depression that persisted through much of the decade. The Liberty statue project was not the only such undertaking that had difficulty raising money: construction of the obelisk later known as the Washington Monument sometimes stalled for years; it would ultimately take over three-and-a-half decades to complete.[60] There was criticism both of Bartholdi's statue and of the fact that the gift required Americans to foot the bill for the pedestal. In the years following the Civil War, most Americans preferred realistic artworks depicting heroes and events from the nation's history, rather than allegorical works like the Liberty statue.[60] There was also a feeling that Americans should design American public works—the selection of Italian-born Constantino Brumidi to decorate the Capitol had provoked intense criticism, even though he was a naturalized U.S. citizen.[61] *Harper's Weekly* declared its wish that "M. Bartholdi and our French cousins had 'gone the whole figure' while they were about it, and given us statue and pedestal at once."[62] *The New York Times* stated that "no true patriot can countenance any such expenditures for bronze females in the present state of our finances."[63] Faced with these criticisms, the American committees took little action for several years.[63]

The foundation of Bartholdi's statue was to be laid inside Fort Wood, a disused army base on Bedloe's Island constructed between 1807 and 1811. Since 1823, it had rarely been used, though during the Civil War, it had served as a recruiting station.[64] The fortifications of the structure were in the shape of an eleven-point star. The statue's foundation and pedestal were aligned so that it would face southeast, greeting ships entering the harbor from the Atlantic Ocean.[65] In 1881, the New York committee commissioned Richard Morris Hunt to design the pedestal. Within months, Hunt submitted a detailed plan, indicating that he expected construction to take about nine months.[66] He proposed a pedestal 114 feet (35 m) in height; faced with money problems, the committee reduced that to 89 feet (27 m).[67]

Hunt's pedestal design contains elements of classical architecture, including Doric portals, and the large mass is fragmented with architectural detail to focus attention on the statue.[67] In form, it is a truncated pyramid, 62 feet (19 m) square at the base and 39.4 feet (12.0 m) at the top. The four sides are identical in appearance. Above the door on each side, there are ten disks upon which Bartholdi proposed to place the coats of arms of the states (between 1876 and 1889, there were 40 U.S. states), although this was not done. Above that, a balcony was placed on each side, framed by pillars. Bartholdi placed an observation platform near the top of the pedestal, above which the statue itself rises.[68] According to author Louis Auchincloss, the pedestal "craggily evokes the power of an ancient Europe over which rises the dominating figure of the Statue of Liberty".[67] The committee hired former army General Charles Pomeroy Stone to oversee the construction work.[69] Construction on the 15-foot (4.6 m) deep foundation began in 1883, and the pedestal's cornerstone was laid in 1884.[66] In Hunt's original conception, the pedestal was to have been made of solid granite. Financial concerns again forced him to revise his plans; the final design called for poured concrete walls, up to 20 feet (6.1 m) thick, faced with granite blocks.[70] [71] The concrete mass was the largest poured to that time.[71]

Fundraising for the statue had begun in 1882. The committee organized a large number of money-raising events.[72] As part of one such effort, an auction of art and manuscripts, poet Emma Lazarus was asked to donate an original work. She initially declined, stating she could not write a poem about a statue. At the time, she was also involved in aiding refugees to New York who had fled anti-Semitic pogroms in eastern Europe. These refugees were forced to live in conditions that the wealthy Lazarus had never experienced. She saw a way to express her empathy for these refugees in terms of the statue.[73] The resulting sonnet, "The New Colossus", including the iconic lines "Give me your tired, your poor/Your huddled masses yearning to breathe free", is uniquely identified with the Statue of Liberty and is inscribed on a plaque in the museum in the base.[74]

Richard Morris Hunt's pedestal under construction in June 1885

Even with these efforts, fundraising lagged. Grover Cleveland, the governor of New York, vetoed a bill to provide $50,000 for the statue project in 1884. An attempt the next year to have Congress provide $100,000, sufficient to complete the project, failed when Democratic representatives would not agree to the appropriation. The New York committee, with only $3,000 in the bank, suspended work on the pedestal. With the project in jeopardy, groups from other American cities, including Boston and Philadelphia, offered to pay the full cost of erecting the statue in return for relocating it.[75] Joseph Pulitzer, publisher of the *World*, a New York newspaper, announced a drive to raise $100,000 (the equivalent of $2.3 million today).[33] Pulitzer pledged to print the name of every contributor, no matter how small the amount given.[76] The drive captured the imagination of New Yorkers, especially when Pulitzer began publishing the notes he received from contributors. "A young girl alone in the world" donated "60 cents, the result of self denial."[77] One donor gave "five cents as a poor office boy's mite toward the Pedestal Fund." A group of children sent a dollar as "the money we saved to go to the circus with."[78] Another dollar was given by a "lonely and very aged woman."[77] Residents of a home for alcoholics in New York's rival city of Brooklyn (the cities would not merge until 1898) donated $15; other drinkers helped out through donation boxes in bars and saloons.[79] A kindergarten class in Davenport, Iowa, mailed the *World* a gift of $1.35.[77]

As the donations flooded in, the committee resumed work on the pedestal.[80] In June, New Yorkers displayed their new-found enthusiasm for the statue, as the French vessel *Isère* arrived with the crates holding the disassembled statue on board. Two hundred thousand people lined the docks and hundreds of boats put to sea to welcome the *Isère*.[81] After five months of daily calls to donate to the statue fund, on August 11, 1885, the *World* announced that $102,000 had been raised from 120,000 donors, and that 80 percent of the total had been received in sums of less than one dollar.[82]

Even with the success of the fund drive, the pedestal was not completed until April 1886. Immediately thereafter, reassembly of the statue began. Eiffel's iron framework was anchored to steel I-beams within the concrete pedestal and assembled.[83] Once this was done, the sections of skin were carefully attached.[84] Due to the width of the pedestal, it was not possible to erect scaffolding, and workers dangled from the armature by ropes while installing the skin sections. Nevertheless, no one died during the construction work.[85] Bartholdi had planned to put floodlights on the torch's balcony to illuminate it; a week before the dedication, the Army Corps of Engineers vetoed the proposal, fearing that ships' pilots passing the statue would be blinded. Instead, Bartholdi cut portholes in the torch (which was covered with gold leaf) and placed the lights inside them.[86] A power plant was installed on the island to light the torch and for other electrical needs.[87] After the skin was completed, renowned landscape architect Frederick Law Olmstead, designer of New York's Central Park and Brooklyn's Prospect Park, supervised a cleanup of Bedloe's Island in anticipation of the dedication.[88]

Dedication

A ceremony of dedication was held on the afternoon of October 28, 1886. President Grover Cleveland, the former New York governor, presided over the event.[89] On the morning of the dedication, a parade was held in New York City; estimates of the number of people who watched it ranged from several hundred thousand to a million. President Cleveland headed the procession, then stood in the reviewing stand to see bands and marchers from across America. General Stone was the grand marshal of the parade. The route began at Madison Square, once the venue for the arm, and proceeded to Battery Park at the southern tip of Manhattan by way of Fifth Avenue and Broadway, with a slight detour so the parade could pass in front of the *World* building on Park Row. As the parade passed the New York Stock Exchange, traders threw ticker tape from the windows, beginning the New York tradition of the ticker-tape parade.[90]

Unveiling of the Statue of Liberty Enlightening the World (1886) by Edward Moran. Oil on canvas. The J. Clarence Davies Collection, Museum of the City of New York.

A nautical parade began at 12:45 p.m., and President Cleveland embarked on a yacht that took him across the harbor to Bedloe's Island for the dedication.[91] De Lesseps made the first speech, on behalf of the French committee, followed by the chairman of the New York committee, Senator William M. Evarts. A French flag draped across the statue's face was to be lowered to unveil the statue at the close of Evarts's speech, but Bartholdi mistook a pause as the conclusion and let the flag fall prematurely. The ensuing cheers put an end to Evarts's address.[90] President Cleveland spoke next, stating that the statue's "stream of light shall pierce the darkness of ignorance and man's oppression until Liberty enlightens the world".[92] Bartholdi, observed near the dais, was called upon to speak, but he refused. Orator Chauncey M. Depew concluded the speechmaking with a lengthy address.[93]

No members of the general public were permitted on the island during the ceremonies, which were reserved entirely for dignitaries. The only females granted access were Bartholdi's wife and de Lesseps's granddaughter; officials stated that they feared women might be injured in the crush of people. The restriction offended area suffragists, who chartered a boat and got as close as they could to the island. The group's leaders made speeches applauding the embodiment of Liberty as a woman and advocating women's right to vote.[92] A scheduled fireworks display was postponed until November 1 because of poor weather.[94]

Shortly after the dedication, the Cleveland *Gazette*, an African American newspaper, suggested that the statue's torch not be lit until the United States became a free nation "in reality":

> "Liberty enlightening the world," indeed! The expression makes us sick. This government is a howling farce. It can not or rather *does not* protect its citizens within its *own* borders. Shove the Bartholdi statue, torch and all, into the ocean until the "liberty" of this country is such as to make it possible for an inoffensive and industrious colored man to earn a respectable living for himself and family, without being ku-kluxed, perhaps murdered, his daughter and wife outraged, and his property destroyed. The idea of the "liberty" of this country "enlightening the world," or even Patagonia, is ridiculous in the extreme.[95]

History since dedication

Lighthouse Board and War Department (1886–1933)

Government poster using the Statue of Liberty to promote the sale of Liberty Bonds

When the torch was illuminated on the evening of the statue's dedication, it produced only a faint gleam, barely visible from Manhattan. The *World* characterized it as "more like a glowworm than a beacon."[87] Bartholdi suggested gilding the statue to increase its ability to reflect light, but this proved too expensive. The United States Lighthouse Board took over the Statue of Liberty in 1887 and pledged to install equipment to enhance the torch's effect; in spite of its efforts, the statue remained virtually invisible at night. When Bartholdi returned to the United States in 1893, he made additional suggestions, all of which proved ineffective. He did successfully lobby for improved lighting within the statue, allowing visitors to better appreciate Eiffel's design.[87] In 1901, President Theodore Roosevelt, once a member of the New York committee, ordered the statue's transfer to the War Department, as it had proved useless as a lighthouse.[96] A unit of the Army Signal Corps was stationed on Bedloe's Island until 1923, after which military police remained there while the island was under military jurisdiction.[97]

The statue rapidly became a landmark. Many immigrants who entered through New York saw it as a welcoming sight. Oral histories of immigrants record their feelings of exhilaration on first viewing the Statue of Liberty. One immigrant who arrived from Greece recalled,

> I saw the Statue of Liberty. And I said to myself, "Lady, you're such a beautiful! [*sic*] You opened your arms and you get all the foreigners here. Give me a chance to prove that I am worth it, to do something, to be someone in America." And always that statue was on my mind.[98]

Originally, the statue was a dull copper color, but shortly after 1900 a green patina, also called verdigris, caused by the oxidation of the copper skin, began to spread. As early as 1902 it was mentioned in the press; by 1906 it had entirely covered the statue.[99] In the belief that the patina was evidence of corrosion, Congress authorized $62,800 to paint the statue both inside and out.[100] There was considerable public protest against the proposed exterior painting.[101] The Army Corps of Engineers studied the patina for any ill effects to the statue and concluded that it protected the skin, "softened the outlines of the Statue and made it beautiful."[102] The statue was painted only on the inside. The Corps of Engineers also installed an elevator to take visitors from the base to the top of the pedestal.[102]

Bedloe's Island in 1927, showing the statue and army buildings. The eleven-pointed walls of Fort Wood, which still form the statue's base, are visible.

On July 30, 1916, during World War I, German saboteurs set off a disastrous explosion on the Black Tom peninsula in Jersey City, New Jersey, in what is now part of Liberty State Park, close to Bedloe's Island. Carloads of dynamite and other explosives that were being sent to Britain and France for their war efforts were detonated, and seven people were killed. The statue sustained minor damage, mostly to the torch-bearing right arm, and was closed for ten days. The cost to repair the statue and buildings on the island was about $100,000. The narrow ascent to the torch was closed for public safety reasons, and it has remained closed ever since.[93]

That same year, Ralph Pulitzer, who had succeeded his father Joseph as publisher of the *World*, began a drive to raise $30,000 for an exterior lighting system to illuminate the statue at night. He claimed over 80,000 contributors but failed to reach the goal. The difference was quietly made up by a gift from a wealthy donor—a fact that was not revealed until 1936. An underwater power cable brought electricity from the mainland and floodlights were placed along the walls of Fort Wood. Gutzon Borglum, who later sculpted Mount Rushmore, redesigned the torch, replacing much of the original copper with stained glass. On December 2, 1916, President Woodrow Wilson pressed the telegraph key that turned on the lights, successfully illuminating the statue.[103]

After the United States entered World War I in 1917, images of the statue were heavily used in both recruitment posters and the Liberty Bond drives that urged American citizens to support the war financially. This impressed upon the public the war's stated purpose—to secure liberty—and served as a reminder that embattled France had given the United States the statue.[104]

In 1924, President Calvin Coolidge used his authority under the Antiquities Act to declare the statue a national monument.[96] The only successful suicide in the statue's history occurred five years later, when a man climbed out of one of the windows in the crown and jumped to his death, glancing off the statue's breast and landing on the base.[105]

Early National Park Service years (1933–1982)

In 1933, President Franklin Roosevelt ordered the statue transferred to the National Park Service (NPS). In 1937, the NPS gained jurisdiction over the rest of Bedloe's Island.[96] With the Army's departure, the NPS began to transform the island into a park.[106] The Works Progress Administration (WPA) demolished most of the old buildings, regraded and reseeded the eastern end of the island, and built granite steps for a new public entrance to the statue from its rear. The WPA also carried out restoration work within the statue, temporarily removing the rays from the statue's halo so their rusted supports could be replaced. Rusted cast-iron steps in the pedestal were replaced with new ones made of reinforced concrete;[107] the upper parts of the stairways within the statue were replaced, as well. Copper sheathing was installed to prevent further damage from rainwater that had been seeping into the pedestal.[108] The statue was closed to the public from May until December 1938.[107]

September 26, 1972: President Richard Nixon visits the statue to open the American Museum of Immigration. The statue's raised right foot is visible, showing that it is depicted moving forward.

During World War II, the statue remained open to visitors, although it was not illuminated at night due to wartime blackouts. It was lit briefly on December 31, 1943, and on D-Day, June 6, 1944, when its lights flashed "dot-dot-dot-dash," the Morse code for V, for victory. New, powerful lighting was installed in 1944–1945, and beginning on V-E Day, the statue was once again illuminated after sunset. The lighting was for only a few hours each evening, and it was not until 1957 that the statue was illuminated every night, all night.[109] In 1946, the interior of the statue within reach of visitors was coated with a special plastic so that graffiti could be washed away.[108]

In 1956, an act of Congress officially renamed Bedloe's Island as Liberty Island, a change advocated by Bartholdi generations earlier. The act also mentioned the efforts to found an American Museum of Immigration on the island, which backers took as federal approval of the project, though the government was slow to grant funds for it.[110] Nearby Ellis Island was made part of the Statue of Liberty National Monument by proclamation of President Lyndon Johnson in 1965.[96] In 1972, the immigration museum, in the statue's base, was finally opened in a ceremony led by President Richard Nixon. The museum's backers never provided it with an endowment to secure its future and it closed in 1991 after the opening of an immigration museum on Ellis Island.[83]

The statue was several times taken over briefly by demonstrators, publicizing causes such as Puerto Rican independence, opposition to abortion, and opposition to US intervention in Grenada. Demonstrations with the permission of the Park Service included a Gay Pride Parade rally and the annual Captive Baltic Nations rally.[111] A powerful new lighting system was installed in advance of the American Bicentennial in 1976. The statue was the focal point for Operation Sail, a regatta of tall ships from all over the world that entered New York Harbor on July 4, 1976, and sailed around Liberty Island.[112] The day concluded with a spectacular display of fireworks near the statue.[113]

Renovation to present (since 1982)

July 4, 1986: First Lady Nancy Reagan (in red) reopens the statue to the public.

The statue was examined in great detail by French and American engineers as part of the planning for its centennial in 1986.[114] In 1982, it was announced that the statue was in need of considerable restoration. Careful study had revealed that the right arm had been improperly attached to the main structure. It was swaying more and more when strong winds blew and there was a significant risk of structural failure. In addition, the head had been installed 2 feet (0.61 m) off center, and one of the rays was wearing a hole in the right arm when the statue moved in the wind. The armature structure was badly corroded, and about two percent of the exterior plates needed to be replaced.[115] Although problems with the armature had been recognized as early as 1936, when cast iron replacements for some of the bars had been installed, much of the corrosion had been hidden by layers of paint applied over the years.[116]

In May 1982, President Ronald Reagan announced the formation of the Statue of Liberty–Ellis Island Centennial Commission, led by Chrysler Corporation chair Lee Iacocca, to raise the funds needed to complete the work.[117] Through its fundraising arm, the Statue of Liberty–Ellis Island Foundation, Inc., the group raised more than $350 million in donations.[118] The Statue of Liberty was one of the earliest beneficiaries of a cause marketing campaign. A 1983 promotion advertised that for each purchase made with an American Express card, the company would contribute one cent to the renovation of the statue. The campaign generated contributions of $1.7 million to the restoration project.[119]

In 1984, the statue was closed to the public for the duration of the renovation. Workers erected scaffolding that obscured the statue from view. Liquid nitrogen was used to remove layers of paint that had been applied to the interior of the copper skin over decades, leaving two layers of coal tar, originally applied to plug leaks and prevent corrosion. Blasting with baking soda powder removed the tar without further damaging the copper.[120] The restorers' work was hampered by the asbestos-based substance that Bartholdi had used (ineffectively, as inspections showed) to prevent galvanic corrosion. Workers within the statue had to wear protective gear, dubbed "moon suits", with self-contained breathing circuits.[121] Larger holes in the copper skin were repaired, and new copper was added where necessary.[122] The replacement skin was taken from a copper rooftop at Bell Labs, which had a patina that closely resembled the statue's; in exchange, the laboratory was provided some of the old copper skin for testing.[123] The torch, found to have been leaking water since the 1916 alterations, was replaced with an exact replica of Bartholdi's unaltered torch.[124] Consideration was given to replacing the arm and shoulder; the National Park Service insisted that they be repaired instead.[125]

The entire armature was replaced. The puddled iron bars used by Eiffel were gradually removed. The new bars that attach to the pylon are made of low-carbon corrosion-resistant stainless steel. The bars that now hold the staples next to the skin are made of Ferralium, an alloy that bends slightly and returns to its original shape as the statue moves.[126] To prevent the ray and arm making contact, the ray was realigned by several degrees.[127] The lighting was again replaced—night-time illumination now comes from metal halide lamps that send beams of light to particular parts of the pedestal or statue, showing off various details.[128] Access to the pedestal, which had been through a nondescript entrance built in the 1960s, was renovated to create a wide opening framed by a set of monumental bronze doors with designs symbolic of the renovation.[129] A modern elevator was installed, allowing handicapped access to the observation area of the pedestal.[130] An emergency elevator was installed within the statue, reaching up to the level of the shoulder.[131]

September 11, 2001: The twin towers of the World Trade Center burn with the Statue of Liberty in the foreground.

July 3–6, 1986, was designated "Liberty Weekend", marking the centennial of the statue and its reopening. President Reagan presided over the rededication, with French President François Mitterrand in attendance. July 4 saw a reprise of Operation Sail,[132] and the statue was reopened to the public on July 5.[133] In Reagan's dedication speech, he stated, "We are the keepers of the flame of liberty; we hold it high for the world to see."[132]

Following the September 11 attacks, the statue and Liberty Island were immediately closed to the public. The island reopened at the end of 2001, while the pedestal and statue remained off-limits. The pedestal reopened in August 2004,[133] but the National Park Service announced that visitors could not safely be given access to the statue due to the difficulty of evacuation in an emergency. The Park Service adhered to that position through the remainder of the Bush administration.[134] New York Congressman Anthony D. Weiner made the statue's reopening a personal crusade.[135] On May 17, 2009, President Barack Obama's Secretary of the Interior, Ken Salazar, announced that as a "special gift" to America, the statue would be reopened to the public as of July 4, but that only a limited number of people would be permitted to ascend to the crown each day.[134] The statue, including the pedestal and base, closed on October 29, 2011 (the day after celebrations to mark the statue's 125th anniversary), for installation of new elevators and staircases, and to bring other facilities, such as restrooms, up to code. Although the statue is closed to the public for up to a year, Liberty Island remains open.[1] [136]

Access and attributes

Location and visiting

The statue is situated in Upper New York Bay on Liberty Island, south of Ellis Island. Both islands were ceded by New York to the federal government in 1800.[137] As agreed in an 1834 compact between New York and New Jersey that set the state border at the bay's midpoint, the original islands remain New York territory despite their location on the New Jersey side of the state line. Land created by reclamation at Ellis is New Jersey territory.[138]

Tourists aboard a Circle Line ferry bound for Liberty Island, June 1973

Entrance to the Statue of Liberty National Monument is free, but there is a charge for the ferry service that all visitors must use, as private boats may not dock at the island. A concession was granted in 2007 to Statue Cruises to operate the transportation and ticketing facilities, replacing Circle Line, which had operated the service since 1953.[139] The ferries, which depart from Liberty State Park in Jersey City and Battery Park in Lower Manhattan, also stop at Ellis Island, making a combined trip possible.[140] All ferry riders are subject to security screening, similar to airport procedures, prior to boarding.[141] The statue, pedestal, and base are presently closed for renovation work.[1] When the site is open, visitors intending to enter the statue's base and pedestal must obtain a complimentary museum/pedestal ticket along with their ferry ticket.[26] Those wishing to climb the staircase within the statue to the crown purchase a special ticket, which may be reserved up to a year in advance. A total of 240 people per day are permitted to ascend: ten per group, three groups per hour. Climbers may bring only medication and cameras—lockers are provided for other items—and must undergo a second security screening.[26]

Inscriptions, plaques, and dedications

Plaque honoring poet Emma Lazarus, with the text of "The New Colossus"

There are several plaques and dedicatory tablets on or near the Statue of Liberty. A plaque on the copper just under the figure in front declares that it is a colossal statue representing Liberty, designed by Bartholdi and built by the Paris firm of Gaget, Gauthier et Cie (*Cie* is the French abbreviation analogous to *Co.*). A presentation tablet, also bearing Bartholdi's name, declares the statue to be a gift from the people of the Republic of France that honors "the Alliance of the two Nations in achieving the Independence of the United States of America and attests their abiding friendship."[142] There is a tablet placed by the New York committee that commemorates the fundraising done to build the pedestal. The cornerstone also bears a plaque placed by the Freemasons.[142]

In 1903, a bronze tablet that bears the text of "The New Colossus" and commemorates Emma Lazarus was presented by friends of the poet. Until the 1986 renovation, it was mounted inside the pedestal; today it resides in the Statue of Liberty Museum in the base. It is accompanied by a tablet given by the Emma Lazarus Commemorative Committee in 1977, celebrating the poet's life.[142]

A group of statues stands at the western end of the island, honoring those closely associated with the Statue of Liberty. Two Americans—Pulitzer and Lazarus—and three Frenchmen—Bartholdi, Laboulaye, and Eiffel—are depicted. They are the work of Maryland sculptor Phillip Ratner.[143]

In 1984, the Statue of Liberty was designated a UNESCO World Heritage Site. The UNESCO "Statement of Significance" describes the statue as a "masterpiece of the human spirit" that "endures as a highly potent symbol—inspiring contemplation, debate and protest—of ideals such as liberty, peace, human rights, abolition of slavery, democracy and opportunity."[144]

Physical characteristics

As viewed from the ground on Liberty Island

Feature[65]	U.S.	Metric
Height of copper statue	151 ft 1 in	46 m
Foundation of pedestal (ground level) to tip of torch	305 ft 1 in	93 m
Heel to top of head	111 ft 1 in	34 m
Length of hand	16 ft 5 in	5 m
Index finger	8 ft 1 in	2.44 m
Circumference at second joint	3 ft 6 in	1.07 m
Head from chin to cranium	17 ft 3 in	5.26 m
Head thickness from ear to ear	10 ft 0 in	3.05 m
Distance across the eye	2 ft 6 in	0.76 m
Length of nose	4 ft 6 in	1.48 m
Right arm length	42 ft 0 in	12.8 m
Right arm greatest thickness	12 ft 0 in	3.66 m
Thickness of waist	35 ft 0 in	10.67 m
Width of mouth	3 ft 0 in	0.91 m
Tablet, length	23 ft 7 in	7.19 m
Tablet, width	13 ft 7 in	4.14 m
Tablet, thickness	2 ft 0 in	0.61 m

Height of pedestal	89 ft 0 in	27.13 m
Height of foundation	65 ft 0 in	19.81 m
Weight of copper used in statue	60,000 pounds	27.22 metric tonnes
Weight of steel used in statue	250,000 pounds	113.4 metric tonnes
Total weight of statue	450,000 pounds	204.1 metric tonnes
Thickness of copper sheeting	3/32 of an inch	2.4 mm

Depictions

A replica of the Statue of Liberty forms part of the exterior decor at the New York-New York Hotel & Casino in Las Vegas, Nevada.

Hundreds of replicas of the Statue of Liberty are displayed worldwide.[145] A smaller version of the statue, one-fourth the height of the original, was given by the American community in Paris to that city. It now stands on the Île aux Cygnes, facing west toward her larger sister.[145] A replica 30 feet (9.1 m) tall stood atop the Liberty Warehouse on West 64th Street in Manhattan for many years;[145] it now resides at the Brooklyn Museum.[146] In a patriotic tribute, the Boy Scouts of America, as part of their Strengthen the Arm of Liberty campaign in 1949–1952, donated about two hundred replicas of the statue, made of stamped copper and 100 inches (2500 mm) in height, to states and municipalities across the United States.[147] Though not a true replica, the statue known as the Goddess of Democracy temporarily erected during the Tiananmen Square protests of 1989 was similarly inspired by French democratic traditions—the sculptors took care to avoid a direct imitation of the Statue of Liberty.[107] Among other recreations of New York City structures, a replica of the statue is part of the exterior of the New York-New York Hotel & Casino in Las Vegas, Nevada.[148]

As an American icon, the Statue of Liberty has been depicted on the country's stamps and coinage. It appeared on commemorative coins issued to mark its 1986 centennial, and on New York's 2001 entry in the state quarters series.[149] An image of the statue was chosen for the American Eagle platinum bullion coins in 1997, and it appears on the reverse, or "tails", side of the currently issued series of Presidential Dollar circulating coins.[22] Two images of the statue's torch appear on the current ten-dollar bill.[150] The statue's intended photographic depiction on a 2010 forever stamp proved instead to be of the replica at the Las Vegas casino.[151]

1971 U.S. airmail stamp depicting the head of the Statue of Liberty

Depictions of the statue have been used by many regional institutions. Between 1986 and 2000, New York State issued license plates featuring the statue.[152] [153] The Women's National Basketball Association's New York Liberty use both the statue's name and its image in their logo, in which the torch's flame doubles as a basketball.[154] The New York Rangers of the National Hockey League depicted the statue's head on their third jersey, beginning in 1997.[155] The National Collegiate Athletic Association's 1996 Men's Basketball Final Four, played at New Jersey's Meadowlands Sports Complex, featured the statue in its logo.[156] </ref>

Reverse side of a Presidential Dollar coin

The statue is a frequent subject in popular culture. In music, it has been evoked to indicate support for American policies, as in Toby Keith's song "Courtesy of the Red, White, & Blue (The Angry American)", and in opposition, appearing on the cover of The Dead Kennedys' album *Bedtime for Democracy*, which protested the Reagan administration's policies.[157] In film, the torch is the setting for the climax of director Alfred Hitchcock's 1942 movie *Saboteur*.[158] The statue makes one of its most famous cinematic appearances in the 1968 picture *Planet of the Apes*, in which it is seen half-buried in sand.[157][159] It is destroyed in the science-fiction films *Independence Day*, *The Day After Tomorrow*, and *Cloverfield*.[157] In Jack Finney's time-travel novel *Time and Again*, the right arm of the statue, on display in the early 1880s in Madison Square Park, plays a crucial role.[160] Robert Holdstock, consulting editor of *The Encyclopedia of Science Fiction*, wondered in 1979,

> Where would [science fiction] be without the Statue of Liberty? For decades it has towered or crumbled above the wastelands of deserted [E]arth—giants have uprooted it, aliens have found it curious ... the symbol of Liberty, of optimism, has become a symbol of science fiction's pessimistic view of the future."[161]

See also

- List of the tallest statues in the United States
- Place des États-Unis
- *The Statue of Liberty*, Ken Burns documentary
- Statue of Responsibility
- Statues and sculptures in New York City

References

[1] "Statue of Liberty National Monument" (http://www.nps.gov/stli/index.htm). National Park Service. 2007-12-31. . Retrieved 2011-10-12.

[2] Schneiderman, R.M. (2010-06-28). "For tourists, Statue of Liberty is nice, but no Forever 21" (http://blogs.wsj.com/metropolis/2010/06/28/for-tourists-statue-of-liberty-is-no-forever-21/). *The Wall Street Journal*. . Retrieved 2011-10-12.

[3] http://whc.unesco.org/en/list/307

[4] National Park Service (1994). *National Register of Historic Places, 1966–1994: Cumulative List Through January 1, 1994*. Washington DC: National Park Service. p. 502. ISBN 0891332545.

[5] "National Monument Proclamations under the Antiquities Act" (http://www.cr.nps.gov/history/hisnps/NPSHistory/monuments.htm). National Park Service. 2003-01-16. . Retrieved 2011-10-12.

[6] "Statue of Liberty National Monument" (http://www.neighborhoodpreservationcenter.org/db/bb_files/STATUE-OF-LIBERTY.pdf) (PDF). New York City Landmarks Preservation Commission. September 14, 1976. . Retrieved 2011-10-12.

[7] Harris 1985, p. 7.

[8] Harris 1985, pp. 7–8.

[9] Khan 2010, pp. 60–61.

[10] Moreno 2000, pp. 39–40.

[11] Harris 1985, pp. 12–13.

[12] Khan 2010, pp. 102–103.

[13] Harris 1985, pp. 16–17.

[14] Khan 2010, p. 85.

[15] Harris 1985, pp. 10–11.

[16] Sutherland 2003, pp. 17–19.

[17] Bodnar, John (2006). "Monuments and Morals: The Nationalization of Civic Instruction" (http://books.google.com/?id=Drhn7XY1WrYC&lpg=PA212&dq=delacroix statue of liberty&pg=PA213#v=onepage&q=delacroix statue of liberty). In Warren,

Donald R.; Patrick, John J.. *Civic and Moral Learning in America*. New York: Macmillan. pp. 212–214. ISBN 1403973962. .
[18] Khan 2010, pp. 96–97.
[19] Khan 2010, pp. 105–108.
[20] Blume, Mary (2004-07-16). "The French icon Marianne à la mode" (http://www.nytimes.com/2004/07/16/style/16iht-blume_ed3_.html). *The New York Times*. . Retrieved 2011-10-12.
[21] "Get the Facts (Frequently Asked Questions about the Statue of Liberty)" (http://www.nps.gov/stli/planyourvisit/get-the-facts.htm). *Statue of Liberty*. National Park Service. . Retrieved 2011-10-19.
[22] "Lady Liberty Reverse Statue of Liberty (1886)" (http://www.usmint.gov/mint_programs/$1coin/?flash=yes&action=reverse). *Presidential $1 coin*. United States Mint. . Retrieved 2010-07-29.
[23] Moreno 2000, pp. 52–53, 55, 87.
[24] Bartholdi, Frédéric (1885). *The Statue of Liberty Enlightening the World* (http://books.google.com/?id=p02VNP45RdsC&dq=pedestal statue of liberty hunt&pg=PP6#v=onepage&q=pedestal statue of liberty hunt). New York: North American Review. p. 42. .
[25] Khan 2010, pp. 108–111.
[26] "Frequently asked questions" (http://www.nps.gov/stli/faqs.htm). *Statue of Liberty National Monument*. National Park Service. . Retrieved 2010-08-10.
[27] Khan 2010, pp. 118, 125.
[28] Harris 1985, p. 26.
[29] Khan 2010, p. 120.
[30] Khan 2010, p. 121.
[31] Khan 2010, pp. 123–125.
[32] Harris 1985, pp. 44–45.
[33] "Seven Ways to Compute the Relative Value of a U.S. Dollar Amount, 1774 to Present/" (http://www.measuringworth.com/uscompare/result.php?use[]=DOLLAR&use[]=GDPDEFLATION&use[]=VCB&use[]=UNSKILLED&use[]=MANCOMP&use[]=NOMGDPCP&use[]=NOMINALGDP&year_source=1875&amount=16000&year_result=2009/). Measuringworth. . Retrieved 2011-10-20. (Consumer price index)
[34] Khan 2010, p. 137.
[35] "News of Norway, Issue 4" (http://www.norway.org/ARCHIVE/News/archive/1999/199904copper/). norway.org. 1999. . Retrieved 2010-07-29.
[36] "Answers about the Statue of Liberty, Part 2" (http://cityroom.blogs.nytimes.com/2009/07/02/answers-about-the-statue-of-liberty-part-2/?scp=7&sq=goddess of democracy statue of liberty&st=cse). *The New York Times*. 2009-07-02. . Retrieved 2011-10-12.
[37] Khan 2010, pp. 126–128.
[38] Bell & Abrams 1984, p. 25.
[39] Bell & Abrams 1984, p. 26.
[40] Khan 2010, p. 130.
[41] Harris 1985, p. 49.
[42] Khan 2010, p. 134.
[43] Bell & Abrams 1984, p. 30.
[44] Moreno 2000, p. 94.
[45] Khan 2010, p. 135.
[46] Bell & Abrams 1984, p. 32.
[47] Khan 2010, pp. 136–137.
[48] Khan 2010, pp. 139–143.
[49] Harris 1985, p. 30.
[50] Khan 2010, p. 144.
[51] Harris 1985, p. 33.
[52] Harris 1985, p. 32.
[53] Harris 1985, p. 34.
[54] "Statue of Liberty" (http://www.pbs.org/kenburns/statueofliberty/timeline/). pbs.org. . Retrieved 2011-10-19.
[55] Harris 1985, pp. 36–38.
[56] Harris 1985, p. 39.
[57] Harris 1985, p. 38.
[58] Bell & Abrams 1984, p. 37.
[59] Bell & Abrams 1984, p. 38.
[60] Khan 2010, pp. 159–160.
[61] Khan 2010, p. 163.
[62] Khan 2010, p. 161.
[63] Khan 2010, p. 160.
[64] Moreno 2000, p. 91.

[65] "Statistics" (http://www.nps.gov/stli/historyculture/statue-statistics.htm). *Statue of Liberty*. National Park Service. 2006-08-16. . Retrieved 2011-10-19.

[66] Khan 2010, p. 169.

[67] Auchincloss, Louis (1986-05-12). "Liberty: Building on the Past" (http://books.google.com/?id=uuYCAAAAMBAJ&lpg=PA86&dq=pedestal statue of liberty hunt design&pg=PA87#v=onepage&q=pedestal statue of liberty hunt design). *New York*: 87. . Retrieved 2011-10-19.

[68] Bartholdi, Frédéric (1885). *The Statue of Liberty Enlightening the World* (http://books.google.com/?id=p02VNP45RdsC&dq=pedestal statue of liberty hunt&pg=PP6#v=onepage&q=pedestal statue of liberty hunt). New York: North American Review. p. 62. .

[69] Harris 1985, pp. 71–72.

[70] Sutherland 2003, pp. 49–50.

[71] Moreno 2000, pp. 184–186.

[72] Khan 2010, pp. 163–164.

[73] Khan 2010, pp. 165–166.

[74] Moreno 2000, pp. 172–175.

[75] Levine, Benjamin; Story, Isabelle F. (1961). "Statue of Liberty" (http://www.libertystatepark.org/statueofliberty/sol3.shtml). National Park Service. . Retrieved 2011-10-19.

[76] Bell & Abrams 1984, pp. 40–41.

[77] Harris 1985, p. 105.

[78] Sutherland 2003, p. 51.

[79] Harris 1985, p. 107.

[80] Harris 1985, pp. 110–111.

[81] Harris 1985, p. 112.

[82] Harris 1985, p. 114.

[83] Moreno 2000, p. 19.

[84] Bell & Abrams 1984, p. 49.

[85] Moreno 2000, p. 64.

[86] Hayden & Despont 1986, p. 36.

[87] Harris 1985, pp. 133–134.

[88] Moreno 2000, p. 65.

[89] Khan 2010, p. 176.

[90] Khan 2010, pp. 177–178.

[91] Bell & Abrams 1984, p. 52.

[92] Harris 1985, p. 127.

[93] Moreno 2000, p. 71.

[94] Harris 1985, p. 128.

[95] "Postponing Bartholdi's statue until there is liberty for colored as well." (http://dbs.ohiohistory.org/africanam/page.cfm?ID=15137). *The Cleveland* Gazette (Cleveland, Ohio): p. 2. 1886-11-27. . Retrieved 2011-10-19.

[96] Moreno 2000, p. 41.

[97] Moreno 2000, p. 24.

[98] Sutherland 2003, p. 78.

[99] "Answers about the Statue of Liberty" (http://cityroom.blogs.nytimes.com/2009/07/01/answers-about-the-statue-of-liberty/). *City Room*. 2009-07-01. . Retrieved 2011-10-19.

[100] "To paint Miss Liberty" (http://query.nytimes.com/gst/abstract.html?res=9D03E0DE143DE433A2575AC1A9619C946797D6CF&scp=2&sq=paint+statue+of+liberty&st=p). *The New York Times*: p. 1. 1906-07-19. . Retrieved 2011-10-19.

[101] "How shall "Miss Liberty"'s toilet be made?" (http://query.nytimes.com/gst/abstract.html?res=9F04E6DE1F3EE733A2575AC2A9619C946797D6CF&scp=3&sq=paint+statue+of+liberty&st=p). *The New York Times*: pp. SM2. 1906-07-29. . Retrieved 2011-10-19.

[102] Harris 1985, p. 168.

[103] Harris 1985, pp. 136–139.

[104] Moreno 2000, pp. 148–151.

[105] Harris 1985, p. 147.

[106] Moreno 2000, p. 136.

[107] Moreno 2000, p. 202.

[108] Harris 1985, p. 169.

[109] Harris 1985, pp. 141–143.

[110] Moreno 2000, pp. 147–148.

[111] Moreno 2000, pp. 72–73.

[112] Harris 1985, p. 143.

[113] Moreno 2000, p. 20.

[114] Harris 1985, p. 165.
[115] Harris 1985, pp. 169–171.
[116] Hayden & Despont 1986, p. 38.
[117] Moreno 2000, pp. 204–205.
[118] Moreno 2000, pp. 216–218.
[119] Daw, Jocelyne (March 2006). *Cause Marketing for Nonprofits: Partner for Purpose, Passion, and Profits*. Hoboken, NJ: John Wiley & Sons. p. 4. ISBN 9780471717508.
[120] Hayden & Despont 1986, p. 81.
[121] Hayden & Despont 1986, p. 76.
[122] Hayden & Despont 1986, p. 55.
[123] Harris 1985, p. 172.
[124] Hayden & Despont 1986, p. 153.
[125] Hayden & Despont 1986, p. 75.
[126] Hayden & Despont 1986, pp. 74–76.
[127] Hayden & Despont 1986, p. 57.
[128] Moreno 2000, p. 153.
[129] Hayden & Despont 1986, p. 71.
[130] Hayden & Despont 1986, p. 84.
[131] Hayden & Despont 1986, p. 88.
[132] Sutherland 2003, p. 106.
[133] "History and Culture" (http://www.nps.gov/stli/historyculture/index.htm). *Statue of Liberty*. National Park Service. . Retrieved 2011-10-20.
[134] Chan, Sewell (2009-05-08). "Statue of Liberty's Crown Will Reopen July 4" (http://cityroom.blogs.nytimes.com/2009/05/08/statue-of-libertys-crown-will-reopen-july-4/). *The New York Times*. . Retrieved 2011-10-20.
[135] Neuman, William (2007-07-05). "Congress to Ask Why Miss Liberty's Crown is Still Closed to Visitors" (http://www.nytimes.com/2007/07/05/nyregion/05liberty.html). *The New York Times*. . Retrieved 2011-10-20.
[136] Raja, Nina (2010-08-10). "Liberty Island to remain open during statue's renovation" (http://www.cnn.com/2011/US/08/10/new.york.liberty.renovation/). CNN. . Retrieved 2011-10-20.
[137] "Early History of Bedloe's Island" (http://www.nps.gov/history/history/online_books/hh/11/hh11m.htm). *Statue of Liberty Historical Handbook*. National Park Service. . Retrieved 2011-10-20.
[138] Supreme Court of the United States (1998). "NEW JERSEY v. NEW YORK 523 U.S. 767" (http://supreme.justia.com/us/523/767/index.html). . Retrieved 2011-10-20.
[139] Ramirez, Anthony, Anthony (June 29, 2007). "Circle Line Loses Pact for Ferries to Liberty Island" (http://www.nytimes.com/2007/06/29/nyregion/29ferry.html?pagewanted=print). *New York Times*. . Retrieved 2011-10-20
[140] "NPS: Liberty and Ellis Island ferry map" (http://www.nps.gov/stli/planyourvisit/ferry-system-map.htm). *Ferry Map*. National Park Service. . Retrieved 2011-10-20.
[141] "For Your Safety and Security" (http://www.nps.gov/stli/planyourvisit/safety-and-security.htm). *Statue of Liberty*. National Park Service. . Retrieved 2011-08-30.
[142] Moreno 2000, pp. 222–223.
[143] Harris 1985, p. 163.
[144] "Statue of Liberty" (http://whc.unesco.org/en/list/307). *World Heritage*. UNESCO. . Retrieved 2011-10-20.
[145] Moreno 2000, pp. 200–201.
[146] "Collections: American Art: Replica of the Statue of Liberty, from Liberty Storage & Warehouse, 43–47 West 64th Street, NYC" (https://www.brooklynmuseum.org/opencollection/objects/163578/Replica_of_the_Statue_of_Liberty_from_Liberty_Storage__and__Warehouse_43-47_West_64th_Street_NYC). Brooklyn Museum. . Retrieved 2011-10-20.
[147] Attoun, Marti (October 2007). "Little Sisters of Liberty" (http://www.scoutingmagazine.org/issues/0710/d-wwas.html). *Scouting*. . Retrieved 2010-08-01.
[148] Goldberger, Paul (1997-01-15). "New York-New York, it's a Las Vegas town" (http://www.nytimes.com/1997/01/15/nyregion/new-york-new-york-it-s-a-las-vegas-town.html?scp=7&sq=new york casino statue of liberty&st=cse). *The New York Times* (New York). . Retrieved 2011-10-20.
[149] "Statue of Liberty postage stamps" (http://www.statueofliberty.org/Statue_of_Liberty_Postage_Stamps.html). Statue of Liberty-Ellis Island Foundation, Inc.. . Retrieved 2011-10-20.
[150] "The redesigned $10 note" (http://www.newmoney.gov/currency/10.htm). *newmoney.gov*. Bureau of Engraving and Printing. . Retrieved 2011-10-20.
[151] Severson, Kim; Healey, Matthew (2011-04-14). "This Lady Liberty is a Las Vegas teenager" (http://www.nytimes.com/2011/04/15/us/15stamp.html?ref=nyregion). *The New York Times* (New York). . Retrieved 2011-10-20.
[152] "State to start issuing new license plates July 1" (http://www.nytimes.com/1986/01/24/nyregion/new-york-to-start-issuing-new-license-plates-july-1.html?scp=3&sq=statue of liberty license plates&st=cse). *The New York Times* (New

York). 1986-01-24. . Retrieved 2011-10-20.

[153] "State license plates to get new look" (http://www.nytimes.com/2000/01/11/nyregion/state-license-plates-to-get-new-look.html?scp=1&sq=statue of liberty license plates&st=cse). *The New York Times* (New York). 2000-01-11. . Retrieved 2011-10-20.

[154] "'Liberty' for New York club" (http://www.nytimes.com/1997/02/14/sports/liberty-for-new-york-club.html?scp=2&sq="new york liberty" statue&st=cse). *The New York Times* (New York). 1997-02-14. . Retrieved 2011-10-20.

[155] Lapointe, Joe (1997-01-12). "Lady Liberty laces up at the Garden" (http://www.nytimes.com/1997/01/12/sports/lady-liberty-laces-up-at-the-garden.html?scp=1&sq=rangers third jersey statue&st=cse). *The New York Times* (New York). . Retrieved 2011-10-20.

[156] Sandomir, Richard (1996-03-29). "Final Four: States put aside their rivalry and try a little cooperation" (http://www.nytimes.com/1996/03/29/sports/final-four-states-put-aside-their-rivalry-and-try-a-little-cooperation.html?scp=1&sq=1996 final four logo statue liberty&st=cse). *The New York Times* (New York). . Retrieved 2011-10-20.

[157] Morris, Tracy S.. "The Statue of Liberty in Popular Culture" (http://traveltips.usatoday.com/statue-liberty-popular-culture-3615.html). *USA Today*. . Retrieved 2011-10-20.

[158] Spoto, Donald (1983). *The Dark Side of Genius: The Life of Alfred Hitchcock*. New York: Ballantine. pp. 262–263. ISBN 0-345-31462-X.

[159] Greene, Eric; Slotkin, Richard (1998). *Planet of the Apes as American myth: race, politics, and popular culture* (http://books.google.com/?id=ZyIiApfhWngC&dq=statue+of+liberty+planet+of+the+apes). Middletown, CT: Wesleyan University Press. p. 52. ISBN 0819563293. . Retrieved 2011-10-20.

[160] Darrach, Brad (1970-06-26). "The spy who came in from 1882" (http://books.google.com/books?id=v1UEAAAAMBAJ&lpg=PA16&dq=statue of liberty's arm time and again&pg=PA16#v=onepage&q=statue of liberty's arm time and again&f=false). *Life* (New York): p. 16. . Retrieved 2011-10-20.

[161] Nicholls, Peter (1979). *The Encyclopedia of Science Fiction*. St Albans, Herts, UK: Granada Publishing Ltd.. p. 14. ISBN 0-586-05380-8.

Bibliography

- Bell, James B.; Abrams, Richard L. (1984). *In Search of Liberty: The Story of the Statue of Liberty and Ellis Island*. Garden City, N.Y.: Doubleday & Co. ISBN 0-385-19624-5.
- Glassberg, David (2003). "Rethinking the Statue of Liberty:". National Park Service. hdl:2450/678.
- Harris, Jonathan (1985). *A Statue for America: The First 100 Years of the Statue of Liberty*. New York, N.Y.: Four Winds Press (a division of Macmillan Publishing Company). ISBN 0-02-742730-7.
- Hayden, Richard Seth; Despont, Thierry W. (1986). *Restoring the Statue of Liberty*. New York, N.Y.: McGraw-Hill Book Company. ISBN 0-07-027326-X.
- Khan, Yasmin Sabina (2010). *Enlightening the World: The Creation of the Statue of Liberty*. Ithaca, N.Y.: Cornell University Press. ISBN 978-0-8014-4851-5.
- Moreno, Barry (2000). *The Statue of Liberty Encyclopedia*. New York, N.Y.: Simon & Schuster. ISBN 9780738536897.
- Sutherland, Cara A. (2003). *The Statue of Liberty*. New York, N.Y.: Barnes & Noble Books. ISBN 9780760738900.

External links

- Statue of Liberty National Monument (http://www.nps.gov/stli/index.htm)
- Statue of Liberty (http://en.structurae.de/structures/data/index.cfm?ID=s0000068) at *Structurae*
- Statue of Liberty-Ellis Island Foundation (http://statueofliberty.org/)
- "A Giant's Task – Cleaning Statue of Liberty", *Popular Mechanics* (February 1932) (http://books.google.com/books?id=4-EDAAAAMBAJ&pg=PA203&dq=Popular+Mechanics+1931+curtiss&hl=en&ei=cLsATbqFMZTjnQf3tNzlDQ&sa=X&oi=book_result&ct=result&resnum=9&ved=0CEQQ6AEwCDgK#v=onepage&q=Popular Mechanics 1931 curtiss&f=true)

Article Sources and Contributors

Sendzimir_process *Source*: http://en.wikipedia.org/w/index.php?title=Sendzimir_process *Contributors*: Brandmeister (old), GregorB, Grotte, Handsclark, Luckyz, Nk, Rail user, T.Sendzimir.Inc, Wizard191, 2 anonymous edits

Tadeusz_Sendzimir *Source*: http://en.wikipedia.org/w/index.php?title=Tadeusz_Sendzimir *Contributors*: Auric, Brandmeister (old), Camdan, CharlotteWebb, DingirXul, Dr. Dan, Folks at 137, Grimlock, HOT L Baltimore, Hetar, JRWalko, Jac16888, Jaraalbe, Johnpacklambert, Kbdank71, LilHelpa, Mhapon, MichaelMaggs, Plasticup, ShelfSkewed, Silin2005, Squiddy, Syrenab, Visor, Wikidan829, Xorx, 10 anonymous edits

Galvanization *Source*: http://en.wikipedia.org/w/index.php?title=Galvanization *Contributors*: 168..., 16@r, 194.200.130.xxx, Andrewa, Andy Dingley, Anthony Appleyard, Arthena, Barticus88, Bjf, Captain-tucker, Ccromwell, Chazmosis, CieloEstrellado, Conversion script, CountMacula, Cwolfe8229, D99figge, DVdm, Daniel Christensen, Diffusionteki, Dratman, EPO, El Mayimbe, Epbr123, Eranb, Esteffect, Farosdaughter, Fdac, FireMonty, Fluffernutter, Frank, Gaius Cornelius, Galvanizer, Gaurangkumar.88, Goldom, Graham87, Grandnational, Greedyhalibut, H3xx, Hdt83, Hereimworldluvme, Heron, Hmains, Hmsbeagle, Iames, Inthend9, Iridescent, Ironfistofanarchy, J. Finkelstein, Japanese Searobin, Jauerback, Jerzy, Jimarib, Jivecat, JohnBlackburne, JordiG, Jpogi, Knuckles, Lars.h.eriksen, Leonard G., LilHelpa, LocoBurger, Locuteh, Malcolm Farmer, Merbabu, Mikeroodeus, Miles2244, Miquonranger03, Mjlissner, Muhandes, Myanw, NickVeys, Odie5533, OldakQuill, Omegatron, Onore Baka Sama, P199, Pauli133, Petecarney, Peterkingiron, Peterlewis, Pill, Pinethicket, Plenumchamber, Polyparadigm, Randrag, Richard L. Peterson, Rjstott, Rmhermen, Rrudzik, Sayfadeen, SchuminWeb, Scotty1.1, Shaddack, SidP, Smarty9911, Splarka, SteveW, Thumperward, Tlim7882, Tommason, Towerman86, Trackball, Venice, Vsmith, Weedking9898, Wizard191, Xionbox, Yamamoto Ichiro, ZS, Ziggy Sawdust, 188 anonymous edits

Strip_steel *Source*: http://en.wikipedia.org/w/index.php?title=Strip_steel *Contributors*: Andy Dingley, Lightwave22, Maradja, Mehulme, RP459, Wizard191, 1 anonymous edits

Redox *Source*: http://en.wikipedia.org/w/index.php?title=Redox *Contributors*: -js-, 05dalesteyn, 0612, ASarnat, Aaronbrick, Abduladam52, Abstraktn, Acanderson18, ActiveSelective, AdamRetchless, Alteripse, Andy Dingley, AndyBQ, Ann Stouter, Annulus, Antandrus, Arbitrary username, AshPseud, Augenis, Avertist, Az29, Bachrach44, Baldhur, Bandn, Bangahli, Baradur X, Basicdesign, Bcnof, Benjah-bmm27, Bensaccount, Bitjungle, Bmicomp, Bomac, Br77rino, Brad101, Brainyiscool, Brane.Blokar, Brianjd, Brianoflee, Brockert, Bryan Derksen, Cacycle, Calculuslover, Calmer Waters, Calz22123, Can't sleep, clown will eat me, Canterbury Tail, CapitalR, Chemicalinterest, ChicXulub, Chipotle, Chrishmt0423, Christian75, Coldplayporta, Cool3, Corrigen, Csgitler, DMacks, DR04, DV8 2XL, Da2ce7, DabMachine, Danewlosed, Danny B-), Dannyallen123, Dar-Ape, Darrien, David D., Davken1102, Defender of torch, Denisarona, Disneian, DocLB, DocWatson42, DoubleBlue, Drphilharmonic, EatAlbertaBeef, Ebz212, EconoPhysicist, Eleu4444, Emote, Euriditi, Faiven, Fanghong, Ffgamera, Finn Bjørklid, Fir0002, Fishaholic9494, Frecklefoot, Frokor, Fuzzform, Gene Nygaard, Gentgeen, Geoffrey.landis, GermanX, Gilgamesh he, Glenn, Goalie6693, Godfinger, Goodnightmush, Gorank4, Gordonrox24, GraemeL, Grafen, Gökhan, Haein45, HarlandQPitt, Hawkes, Hayabusa future, Heron, Hgrobe, Hugo-cs, Icseaturtles, Ihasbrainsftw, Ike9898, Inky, Iridescent, Itub, Ivannoriel, J G Campbell, JTN, Jackfirst, Jakevoorhees, JidGom, Jimp, Jjotter, Jleedev, Jnestorius, JohnJohn, Johnheritage, JorisvS, Joseph Solis in Australia, JoshuaKuo, Jrockley, Jwink3101, Kanthoney, Katanada, Keegscee, Keenan Pepper, Kilo-Lima, Kilonum, Kingpin13, Kjkolb, KnowledgeOfSelf, Kr5t, La Pianista, Lamro, Lexi Marie, Lfh, LiDaobing, Lilac Soul, Limhes, Loggie, Luna Santin, Lupin, MC10, MER-C, MREGGMAN, Maddie!, Makeemlighter, Mantissa128, Marc Venot, Marsupilamov, Materialscientist, Matthuxtable, Mbimmler, Meeples, Melamed katz, Memodude, Miborovsky, Michael Hardy, Michaeloqu, Mike Rosoft, Mikespedia, Miskatonic, Mitchellc, Mizst, Monkeyjunky, Mormatil, Munita Prasad, Mutinus, Myran4000, Naerii, NawlinWiki, Nescio, Nick C, NickBush24, Nosferatütr, OTB, Ohiostandard, Oleg Alexandrov, Ollie holton, Omegatron, Omer88f, Oncology group, OnePt618, Palaxzorodice, Paleorthid, Paul August, Pavlo Chemist, Pdch, Peacekills, Philip Trueman, Piano non troppo, Pinethicket, Polyamorph, Polyparadigm, Prachi pant, Preeya2112, Prowikipedians, Public Menace, Qatter, Qfl247, Quentin X, R.E. Freak, R3m0t, Radon210, RainbowOfLight, Ranjithsutari, Rd108, Renebeto, Reo On, RexNL, Rfl, Rich Farmbrough, Richard001, Rifleman 82, Righthing, Ristorilsm, Robinatron, ST47, Sagaciousuk, Sazzad Hossen, ScAvenger, Scratchaholiks, Semper discens, Sentynel, Sfahey, Shadowjams, Shanes, Shellreef, Shenme, Shinkolobwe, Sillybilly, Silverhand, Skihatboatbike, Smalljim, Smokeala, Smokefoot, Soba0401, Some jerk on the Internet, Someones life, Sonett72, Soulkeeper, Spoon!, Steinsky, Stevefish, Stickee, SubZero..Vighu.., Sudosiko, Supearnesh, SuperHamster, SuperTycoon, Svdmolen, Swpb, TBadger, Tarquin, Taxman, Tdb123, TedE, Tetracube, The High Fin Sperm Whale, TheGWO, TheGoGetter7, Tide rolls, Tilmitt, TimBentley, TimVickers, Todan, Toh, Tom harrison, Tommy2010, Towerman86, Trainspotter, Trempealeau Pi, Tsemii, Turkeymanwill, TutterMouse, Twisted86, Ubiquita, Una Smith, Unfree, User A1, V8rik, Vectro, Vsmith, WMSwiki, Wavelength, Wereon, Wfgiuliano, Wickey-nl, Wikicrow, Wikipe-tan, WikipedianMarlith, Winetype, Wknight94, Wmahan, WolfmanSF, Woohookitty, Xenowiki, Yamamoto Ichiro, Yettie0711, ZayZayEM, Zkester4, ~K, Јованвб, Александър, محبوب عالم, 703 anonymous edits

Hot-dip_galvanizing *Source*: http://en.wikipedia.org/w/index.php?title=Hot-dip_galvanizing *Contributors*: Ali@gwc.org.uk, Andy Dingley, Ben Boldt, BenFrantzDale, Blanchardb, Canterbury Tail, Cycotic, Daniel Christensen, Diffusionteki, Dogears, Doodle77, Element16, Fru1tbat, Gaurangkumar.88, Helpful4sure, Hqb, Idacoolboy, Iepeulas, L314t, Materialscientist, Maury Markowitz, Michael Devore, Mmxsaro, Mmxx, MrBell, MrOllie, Muscarella, Observer29, Omegatron, P199, Perskyro, Phasmatisnox, Possum, R'n'B, R6144, Randrag, Retodon8, Rettetast, Rifleman 82, Ronhjones, Salmanazar, Securiger, Semperf, Shaddack, Slakr, Stone, The Yeti, Thumperward, Trkiehl, Vadim Makarov, Vgeraskin, Wizard191, York3k2, علی ویی, 101 anonymous edits

Institute_of_Materials,_Minerals_and_Mining *Source*: http://en.wikipedia.org/w/index.php?title=Institute_of_Materials%2C_Minerals_and_Mining *Contributors*: Aaditya 7, AvicAWB, Cheveney, Cj1340, Cutler, Daniel.auger, DinosaursLoveExistence, Dr Gangrene, Dr Sumeet, Edward, Judicatus, Kelapstick, Mervyn, Nunastaniaszek, Peterlewis, Pinoakcourt, Thoughtcat, Tim!, Timrem, Webcontentmanager2, Woohookitty, Wtshymanski, 24 anonymous edits

Ostoja_coat_of_arms *Source*: http://en.wikipedia.org/w/index.php?title=Ostoja_coat_of_arms *Contributors*: Adan Blociszewski, Alhierd, Andreas Philopater, Ar tas, Avalokitesvara, Bihco, Cambridgehistory, Camdan, Colonies Chris, Emax, EoGuy, Gustavo Szwedowski de Korwin, Halibutt, JRWalko, Karybut, Khazar, Koavf, Kobrabones, LeoHerz, Logologist, Martinostoja, Severo, Steve crayon, Wames, 33 anonymous edits

Cross_of_Merit_(Poland) *Source*: http://en.wikipedia.org/w/index.php?title=Cross_of_Merit_%28Poland%29 *Contributors*: Bubpeter, Caerwine, Gracenotes, JRWalko, Kintetsubuffalo, Milton Stanley, MrDolomite, Nihil novi, Pibwl, Robert Prummel, Virgencita, quemequedecomoestoy, Wayland, Witkacy, Xdamr, 1 anonymous edits

Royal_Swedish_Academy_of_Sciences *Source*: http://en.wikipedia.org/w/index.php?title=Royal_Swedish_Academy_of_Sciences *Contributors*: Aaker, Alarm, Andrew Dalby, Avjoska, Bjankuloski06en, Caerwine, CalJW, Carabinieri, Circeus, Complex01, Dahlis, Djr xi, DuncanHill, GcSwRhIc, Giftlite, Hayabusa future, Hede2000, JYOuyang, Jac16888, Jaraalbe, JoergenB, Jóhann Heiðar Árnason, Kaaveh Ahangar, LA2, Liftarn, Marie Poise, Masterpiece2000, Mic, Mitrius, Mkallgren, NYScholar, Njsg, Numbo3, Panda, Pete.Hurd, Pion, PuzzletChung, Reedy, Rsabbatini, Saihtam, Shanes, Skomorokh, Splash, Template namespace initialisation script, Tide rolls, Tomas e, Uppland, Vision Thing, VolatileChemical, Xrisham, Zarvok, 23 anonymous edits

Statue_of_Liberty *Source*: http://en.wikipedia.org/w/index.php?title=Statue_of_Liberty *Contributors*: (jarbarf), .:Alex:., 13thchild, 21655, 22zach919, 28421u2232nfenfcenc, 3.14159265, 3232330, 4twenty42o, 67.harry, 7&6=thirteen, 714s2k, 83d40m, 9jules9, A More Perfect Onion, A2Kafir, AJR, AKtarus, Aandrei, Aaron Brenneman, AbJ32, Abce2, Absurdist, Acalamari, Acomuna, Acroterion, Adam Carr, Adam Keller, Adamfuhrer, Adashiel, Adawg123, Addshore, Adel Montevega, AdnanSa, Ahoerstemeier, Aillema, Aitias, Ajayit302, Ajd, Ajraddatz, Akamad, AlainV, Alan Canon, Alan Liefting, Alansohn, Alaphent, Alborzagros, Aleenf1, Alex 01king, Alex.muller, Alex43223, Alexf, AlexiusHoratius, Alextrevelian 006, Ali@gwc.org.uk, Allemannster, Allissonn, Allstarecho, Allstatemurph, Alphachimp, AlternateAlternateHamburger, Alvinrune, Alvis, Amadscientist, Amandajm, Amicon, Amitch, Amputation, Andafia, Andonic, Andreas Kaganov, Andreas Toth, Andres, Andrewlp1991, Android79, Andy Marchbanks, AndyZ, AngChenrui, AngelOfSadness, Angela, Animum, Anna Frodesiak, Anne McDermott, AnonMoos, Another Believer, Antandrus, Antigrandiose, AntonBreusov, Antonio Lopez, AntonioMartin, Anturiaethwr, Ap58, Apiya27, Appraiser, Aramgutang, Aranel, Archer3, Ark, Arkannis, Arm, Armatura, Arpingstone, Arrowcutter, Artoasis, ArtsCountyFair, Aruton, Asfsdfdf, Asgodbarrantes, Ashley Graham, Aspensti, Astroview120mm, Astuishin, Atlant, AuburnPilot, Aude, Audiohifi, Avala, Avicennasis, Avnjay, AxelBoldt, AxelleD, AzaToth, B, Bad Gopher Gear, Badlin, Bahstahd, BarenakedKevin, Barneca, Barong, Barrettmagic, Barrkel, Barry Moreno, Bastique, Batman2005, Bayang, Bazzaro123, Bbatsell, Bbenjoe, Bcorr, Bedford, Beeerrriiiissss, Before My Ken, BenBerry, BenFrantzDale, Benbest, Benea, Benjdowney, Bennybp, Bensin, Benw, Beowulf922, Berean Hunter, Bernfarr, Bfigura's puppy, Bibliomaniac15, Biersaufer, BigMacSC99, Bigpieofpi, Biker Biker, BillC, Billy bob7823, BillyBobChhist, Binx6693, BirdValiant, Birdhurst, Biruitorul, Bizkit311J, Bjeversole, Blanchardb, BlastOButter42, Blawhaw, Blehfu, Blickblocks, Bllasae, Blood Red Sandman, Blouseman, Bnag, Bob bobato, Bob to the e, Bobber2, Bobblewik, Bobo192, Bobsyouruncle(notreally), Bogey97, Bongbang, Bongomaster, Bongwarrior, Bonzolive, Bookandcoffee, Bookwormrwt, Boomshadow, Borgx, Borisblue, Boshtang, Bpeps, BrOnXbOmBr21, Brad Rousse, Brahmaputra, Branaman, BranderChatfield, Brandmeister (old), Brandon5485, Brholden, Brian0918, Brianboulton, Brianga, Brianhe, Briannalong, Brisvegas, Bronks, Bruguiea, Bryan Derksen, BryanG, Bsadowski1, Bsilverthorn, Bsroiaadn, Bubba hotep, Burntsauce, Buster2058, BusterD, C mon, C.Fred, CLW, CWY2190, CalJW, CalifornyaRIP, Caltas, Caltrop, Calvin 1998, Camelbinky, Campbell-53515, Can't sleep, clown will eat me, Cantiorix, Canton Viaduct, Cantus, Capricorn42, Captain Caveman, Captain Weirdo the Great, Captain Wikify, Carbuncle, Carldubois14, Carly.munyard, Carnildo, Carptrash, Casull, Catgut, Caudex Rax, Cbrittain10, Cburnett, Centuriono, Cflm001, Chaerani, Chancemichaels, Charles, Chem-awb, Chendy, Chensiyuan, Chepry, Chile Nose Jam, Chilliwacker, Chinneeb, Cholmes75, Chopin06, Chowbok, Chris 73, ChrisRuvolo, Chrisbuckingham, Christina Silverman, Christopher Parham, Christyanthemum, Chriswiki, Chuck Carroll, Chupri, Citicat, Cj watson, Cjaysson, Cjs2111, Ckatz, Clair4133, Clarkbhm, Clawed, Clemmy, Cliff smith, CliffC, Closedmouth, Closenplay, Clowfacejoe, Coasterlover1994, Codenamecuckoo, Coelacan, Coemgenus, Coffee, Cometstyles, Computerjoe, Comrade009, Confiteordeo, Conscious, Conti, Contributor777, Conversion script, Cooksi, CoolCameron56, CoolyTubeHR, Cordell, Corporal Tunnel, Corpx, Cosmic Latte, Courcelles, Cp111, Cpl Syx, Cptkrunch13, Cracker dane, CrankyScorpion, Crazyyoda1138, Cremepuff222, Creyes, Cricobr, Crimson Red Fox, Crimson30, Crissov, Csernica, Cst17, Cstuder, Ctrl build, Cutler, Cwswb, Cyclonenim, D, D0762, DARTH SIDIOUS 2, DCGeist, DHN, DJ Clayworth, DMacks, DO'Neil, DRosenbach, DT29, DVD R W, Da500063, Dabomb87, Daboymany, Dac04, Dale Arnett, DaltonCox78, Dalyup!, Damicatz, Damiens.rf, Dan4142, Danceternal, Daniel Case, Danielfolsom, Danny Beaudoin, DannyQuack, Dar-Ape, Dark Mage, Darkcelldarkcell, Darkfrog24, Darknesslover5000, Darkwaterrrr, Dave6, David J Johnson, DavidLevinson, DavidWBrooks, Dawn Bard, Dbfirs, Dcamp314, DeadEyeArrow, Deadjune, Deadkid dk, Dean Wormer, Deanhellene, DearPrudence, Debresser, Decumanus, Deepcore,

Delaywaves, Delldot, Demmy, Den fjättrade ankan, Dennis Brown, Deor, DerHexer, Desmay, Destroyman, Dfloren1, Dgies, Diabeticdude123, Didiercollard, Dimadick, Dirac1933, Dirkbb, Discospinster, Djflem, Djmutex, Dlohcierekim, Dmadeo, Dmitrdmi, Doeboi, DonFB, Doncram, DoubleBlue, Dougweller, Doulos Christos, Dpbsmith, DrSad, DrSmacky, Dralwik, Drat, Dread Specter, Dream out loud, Drummer743, Dschwen, Dspradau, Dspuller, Dthomsen8, Dubaduba, Dudemanfellabra, Duoraven, Dupz, Dureo, Durova, Dust Filter, Dustin Pearson, Dyjodee, Dysepsion, Dúnadan, E Pluribus Anthony, EAi, EBundy, EJF (huggle), EOZyo, ESkog, EastCoastGhost, Econrad, Ed g2s, Edmundbaird, Edward321, Edwardlalone, Edwardopuddleduck, EeepEeep, Eerattler, Egmontaz, Eidimon, Elaborating, Elcobbola, Eldrua, ElinorD, Eliot Stearns, Elipongo, Ellomate, Eloquence, Eluchil404, Emir34, Eniagrom, Enkayache, Entropy, Enviroboy, Eoghanacht, Epbr123, Eric-Wester, ErikNY, ErkDemon, Escape Orbit, Esparcadia, Eu.stefan, Evercat, Everyking, Excirial, Exeunt, Extransit, F.bendik, FF2010, Fabrice Ferrer, Face-2-face, Falcanary, Falcon8765, Falcorian, Fanra, Favonian, Fawzyano, FeanorStar7, Fetchcomms, Fieldday-sunday, FigmentJedi, Fishbert, Flame111, Flauto Dolce, Flcelloguy, Flewis, Flopper89, Flowerpotman, Fluffernutter, Flux.books, Foodman, Footwarrior, Fox, Foxfan24, Foxxy497, Fragileautopsy, Fran Rogers, FrancoGG, Freakmighty, FreplySpang, Fresh2deafbill, Froid, Fuhghettaboutit, FullMetal Falcon, FunPika, Funandtrvl, Furado, Furry, Furrykef, Fvdham, Fyyer, GB fan, GDonato, GaMeRulnEr, Gadfium, Gail, Gaius Cornelius, Galoubet, GamblinMonkey, GaretJax777, Garronb, Gary King, Gazimoff, Gdo01, GeeJo, Geographer, George Ponderevo, George The Dragon, Gfoley4, Ghosts&empties, Giftlite, Gilliam, Giovanni33, Gizzakk, Gjd001, Gjs238, Glabangsta666, Gogo Dodo, Good Olfactory, Goodvac, Gop 24fan, Gopher292, Gorgan almighty, Grafen, Graham87, GrandPoohBah, GreatBritain1843, Greenlantern787, GregU, Greghenderson2006, GregorB, Gregory Arkadin, Greudin, Gro-Tsen, GrooveDog, Ground Zero, Grutness, Gscshoyru, Guillaume75000, Gunit13, Gurch, Gwen Gale, Gwernol, Gyea mo, Gyrofrog, H, HJ Mitchell, Hadal, HaeB, Haham hanuka, Hairy Dude, Halgin, HandsomeFella, Hangfromthefloor, Hansonhess11, Happyme22, Hauganm, Hda3ku, Hdt83, HeartOfGold, HedgeFundBob, Heklim, Helloaday, Hesperian, Hetar, HexaChord, Hfodf, HiDrNick, Hidal, Highland14, Hires an editor, Hlucho, Hmains, Hmrox, Hobartimus, HolyBaloney, Honeyman, HoodedMan, Hotbananaomelets, Hotcrocodile, Hqb, Hsanchia, Hu12, Hudders007, Hugh Manatee, Humus sapiens, Husond, Hydrogen Iodide, I already forgot, I-10, INaNimAtE, IRP, ITOMIC, ITim501, Ian Pitchford, Ian.thomson, Ida Shaw, Idont Havaname, Igordebraga, Ilikemenalot, Ilkg952315, Illnab1024, Imaglang, Iman&Sofia, Imran, Indu Singh, Ingolfson, Ingsoc, Inktvis, Insanity Incarnate, InsertYourFaceHere, Intelligentsium, Ioscius, Ipatrol, Iridescent, IsildursBane, Islander, Ithakiboy, Ivysaur, Iwsh, Ixfd64, J.Gowers, J.delanoy, JForget, JJay, JMS Old Al, JNW, JPX7, JQQ, JV Smithy, JW1805, Ja 62, Jab843, Jack1956, Jackehammond, JackofOz, Jacooks, Jaerik, Jake Wartenberg, Jakedeut12, James086, James100, JamesMLane, Jameslwoodward, Jannizz, Jared, Jarhed, Jason Palpatine, JasonAQuest, Jasonwilkes42, Javert, Jaxl, JayJasper, Jayen466, Jcrook1987, Jd4508, Jebba, Jeff G., Jeffec, Jeffrey Mall, Jennavecia, Jennica, Jerry7171, Jersey1324, JerseyDevil, Jeshii, Jess567, Jh51681, Jhoffman13, Jhsounds, Jibran1, JimMillerJr, Jivecat, Jjasi, Jklin, Jleon, Jllm06, Jmlk17, Jnelson09, Jnewmano, JoJan, Joaquín Martínez Rosado, Joelvanatta, Joey80, Joeyconnick, Johhny-turbo, John, John254, Johnbod, Johnnynight, Jojhutton, Jon Rob, Jonas brother lover(joe), Jons63, Jonterry4, Joseph Solis in Australia, Joseph398, Josephf, Jossi, Jpbowen, Jpgordon, Jpk, Jrett060, JruASAP, Judicatus, Juhachi, Juju73700, Jujubee96, Jujutacular, Juliancolton, Jusdafax, JustAGal, K10wnsta, K1Bond007, KConWiki, KFan II, Kaaveh Ahangar, Kafziel, Kaicarver, Kaiser599, Kaisershatner, Kakofonous, Kannie, Kansaisamurai, Karl Dickman, Kasreyn, Katalaveno, Katharynm, Katieh5584, Kavibhalla, Kayau, Kaypoh, Kbdank71, Keegan, Keilana, Kelbag, Kelisi, Kelseybear1997, Kenyon, Kessler, Kevin Hanse, KevinTR, Khan singh, Kidlittle, Killiondude, Kilo-Lima, King of Hearts, Kingpin1055, Kingpin13, Kingturtle, Kittykt495, Kku, Kkxfer, Klundarr, KnowledgeOfSelf, Kohdok, Kokin, Konstable, Korovioff, Kostisl, Kotiwalo, Kowloonese, KrakatoaKatie, Krator, Krenn, Kubigula, Kukini, Kuru, Kweku6, L Kensington, LAX, LFaraone, LOL, La Pianista, LaFoiblesse, LachlanA, Lacrimosus, Lalaland1000, Landon1980, LarryGilbert, Latitude0116, Latitudinarian, Ld100, LeMarsu, LeRoi, Lee Daniel Crocker, Leifern, Lemonade100, Lesmisgirl10, Less Than Math, Leszek Jańczuk, Levineps, Lewissmith05, LibRocks, Lightmouse, Lightsup55, Ligulem, Lil'Meth, LilWille, LisaSandford, Lockesdonkey, Lomn, Loodog, Looslion, Lord David, Lordmetroid, Lou.weird, Lovelac7, LovesMacs, Lowellian, Lpgeffen, Lucy McGough, Ludwigpk, Lugnuts, Luk, Lukep913, Lukewilbur, Lumaga, Luna Santin, Lunchscale, Lupin, Lupo, M.nelson, M2K 2, MER-C, MONGO, MPerel, MacGyverMagic, MackSalmon, Macslacker, Madchester, Madhero88, Madkin999, Maerk, Magic Lemur, Magicalex9, Magog the Ogre, Maias, Maikel, Majorly, Malinaccier, Mandarax, Mangwanani, Mannafredo, Manwe, Mapetite526, Marek69, Mark, MarkD19707, MarmadukePercy, MaroonZ24, MarsRover, Marsbound2024, Martin451, Martinvl, Masciare, MassKnowledgeLearner, Master of Puppets, Master son, Master50, Masterknighted, Masterpiece2000, Matchups, Mateo SA, MattCox78, Mattgirling, Matthiggins69, Mattman4593, Mattworld, Mav, Max Naylor, MaxVeers, Maxis ftw, Mbc362, Mbecker, Mboverload, Mbxp, Mccragus, Mcoupal, Mdeeds, Megaman en m, Melaen, Menchi, Merope, Metasquares, MetsFan76, Mgiganteus1, Mhacdebhandia, Mhking, Michael Dorosh, Michael Glass, Michael Greiner, Mikeyc, MinorProphet, Miquonranger03, Miranda, Misterweiss, Mjfield, Mjmemm, Mkdw, Mochi, Modster, Modulatum, Mondarats, Monkey Bounce, Monkey66, Monre, Montrealais, Moonraker0022, Mortyman, Moshe Constantine Hassan Al-Silverburg, Moulder, Mouloud47, Mouse is back, Mr. Vernon, MrD9, Mrcoyote123, Mrmaroon25, Mrprez27, Mrschimpf, Ms2ger, Msbarker, Mschlindwein, Mtd2006, Muckish, Mufka, Mugsywwiii, Mukanil, Mulad, MusicMaker5376, Mustaflickertafin, Myanw, Mygerardromance, Myrabella, Mythdon, MzVEEx3, N96, Nakon, Naohiro19, Narom, Natalie Erin, Natural RX, NawlinWiki, Nbtumkdfghfgh, Neckro, NeilFraser, NellieBly, NeoChaosX, NeoUrfahraner, Nepat14, NerdyScienceDude, Nerothanes, Nesp0rts11111, Neutrality, Nev1, Neverquick, Nevit, New Yorker 763, NewEnglandYankee, Newone, Newsonline12, Nharkrader, Nicholas.goder, Nick, Nick Number, Nick Ottery, Nick45726, Nicksgigolo, Nightscream, Nihiltres, Niki2006, Nikkimaria, Njohns19, No Guru, Noctibus, Norm mit, Northofdc, NotAnonymous0, Notinasnaid, Nowa, Npgallery, Ntsimp, Nunh-huh, Nutiketaiel, Nxclass, Nyarkni, Nygoodliving, ONEunited4ever, Oben, Obsolium, Ocaasi, Ocatecir, Ocean Shores, OddibeKerfeld, Ohconfucius, Ohnoitsjamie, OlEnglish, Olivier, Omnibus, One, One more night, Only, OptimumPx, Orange Suede Sofa, OrbitOne, Orbst, Orchidor, Osoflase195, Ospalh, Over The Desk, Owen, OwenX, Oxymoron83, P0lyglut, P3Pp3r, PMBO, Paine Ellsworth, Pairadox, Paranoid, Parhamr, Paris By Night, Parklands cobbler, Party, PatGallacher, Patrick, Patrickneil, Paul Klenk, Paul Soth, PaulHanson, Paulinho28, Paulj032, Paulmewis, Pb30, Pderuvo, Pedro, Pengyanan, Pennbradly, Pennywisdom2099, Pepper, Per Honor et Gloria, Peripitus, Persi r, Persian Poet Gal, Peruvianllama, Peter AUDEMG, PeterHuntington, PeterSymonds, Peterpipper, Petersam, Pethan, Pettiebone, Pharos, Pharzo, Phearson, PhilSC, Phileas, Philip Stevens, Philip Trueman, PhilipO, Philwelch, Phoenixrod, Piano non troppo, Pictureuploader, Pigboot, Piledhigheranddeeper, Pinethicket, Pinkadelica, Pioneer-12, Pirateer, Piroroadkill, Pixel, Pjb dinky, Pjbflynn, Plasticbadge, Plastikspork, Pmj, Pnh, Poetic Decay, Polluxian, Pooping pants, Postdlf, Prari, Prescott, Professional106, ProhibitOnions, Prolog, Promethean, Proxima Centauri, Prylon, Pschemp, Public Menace, Puchiko, Pumpkingrrl, PunkdPanther, Pyrofork, Quasipalm, Quisquillian, Qxz, Qyd, R'n'B, RRKennison, RTexasUSA, Rachellovesmusic, Radical Mallard, RadicalPi, Radon210, Raelx, Ragesoss, Random user 39849958, RandomLurker, RandomP, Raptus Regaliter Cattus Petasatus, Rasnaboy, Raven4x4x, Raysonho, Razorflame, Rbakker99, Rdikeman, Reach Out to the Truth, Reddi, Redthoreau, Redtigerxyz, Rees11, Remember the dot, Renata3, Retran, Rettetast, RexNL, Reywas92, Rfc1394, Rgoodermote, Rhialto, Rich Farmbrough, Richard Arthur Norton (1958-), Richard D. LeCour, Richard Harvey, Richie, Richiekim, Ricimer, Ridge Runner, Rigoureux, Rjensen, Rjwilmsi, Robearsn, Robert K S, RobertG, Robertdorsky, Robin klein, RogerGreenhalgh, Roke, Roodog2k, Root Beers, Rouge2, RoyBoy, RoySmith, Royut, Rrburke, Rreagan007, Rror, Rumping, RunOrDie, RussNelson, Rustinonthevine, Rusty1111, Ryulong, Ryz05, SDC, SMC, Saayiit, Sadads, Sadpwner, Salamurai, Salvio giuliano, Sam Korn, Samwb123, SandyGeorgia, Sannse, Santryl, Sappy123110, SarahStierch, Saric, Saros136, Sarsaparilla39, Scapler, Scewing, SchfiftyThree, SchnappM, SchuminWeb, Science4sail, Scjessey, Scooby42, Scott Mingus, Scotthatton, SeNeKa, Seabhcan, Sean WI, Sean7341, Seaphoto, Seba5618, SebasTorrente, Seduisant, Sekicho, Selket, Senator Palpatine, Servant1, Seth Ilys, Seth ze, Sfahey, Shadowjams, Shafei, Shaka, ShakataGaNai, Shalom Yechiel, Sharkface217, Shell Kinney, Sherreyes, Shinmawa, Shoreranger, Shyam, Siebren, Silly Dan, Silverblaster, Silverdaemonskye, Simeon, Sintaku, Sirina, Sjakkalle, Skapur, Skidude9950, Skier Dude, Skitnoon, Skomorokh, Sky83, SkyWalker, Skysmith, Slambo, Sln3412, Slon02, Smallbones, Smkyle1, Snigbrook, Snortonfire, Snowmanradio, Snowolf, Socal gal at heart, Socob, Soliloquial, Some jerk on the Internet, Someguy1221, Sooner Dave, SorryGuy, Sottolacqua, Soupa frk, Soupnyc807, Soy gil2004, SpK, SpaceFlight89, Spam, Sparkit, Special-T, Spencer, Spettro9, Spotteddogsdotorg, SpuriousQ, Spymanut, Sramz22, Ssteedman, St07, StAnselm, Stantonjones, Starsword333, Station1, Statuelibertylives, Staxringold, Stefano KALB, Stemonitis, Stephen, StephenBuxton, Stephenb, Sterling Westland, Steven Zhang, Stevepoppers, Stimpy9337, Stoical, Storm Rider, Sturm55, Stymphal, Suffusion of Yellow, SummerPhD, SunshineofYourLove, Supersam499, Supspirit, Surgeonsmate, Sus scrofa, Swedish fusilier, Swid, Swtpc6800, T saston, T-borg, TMC1982, Tabletop, Tanglewood4, Tangotango, Tanis573, Taurrandir, Tbhotch, Tcncv, Ted87, Tedickey, Tempodivalse, TenPoundHammer, Terence, Th1nk ur b4d, The Duke of Waltham, The Holy Roman Emperor, The Legendary Ranger, The Mystery Man, The PIPE, The Rambling Man, The Show-Ender, The Thing That Should Not Be, The undertow, The wub, TheFearedMachina, TheHYPO, ThePedanticPrick, Thebird25, Thebob000, Thegoofyboy, Theo F, Theresa knott, Therustinator132, Thewetrat, Thingg, Thinking of England, Thisisbossi, Thismightbezach, Thom the man, ThreeBlindMice, Thryduulf, Thumperward, Tiddly Tom, Tide rolls, TimSPC, Time3000, Timjarrett, Timotheus Canens, Timt1006, Titoxd, Toddst1, Tom, Tombomp, Tommy2010, Tone, Tony Corsini, Tony1, Tpbradbury, Transpoman, Tregoweth, Trevor MacInnis, Treybien, Trovatore, True Pagan Warrior, Trusilver, Trödel, Ttownfeen, Turian, Twaz, Twinkler4, Twothousandandnine, Tysto, UberMan5000, Ukexpat, Ulric1313, UltimaRatio, Ultraviolet scissor flame, Uncle Dick, Unclemikejb, UpstateNYer, Uris, Ursasapien, Useight, User86654, Utcursch, Vanished User 4517, Vanka5, Vegaswikian, Versus22, Vesperholly, Vgranucci, Vicki Rosenzweig, Victor Chmara, Vilerage, Vipinhari, Viriditas, Vivio Testarossa, VolatileChemical, Volland, Voyager, Voyagerfan5761, Vranak, Vrenator, Vsmith, Vwenzel, Vzbs34, WJBscribe, WTGDMan1986, Wackywace, Wafulz, Waggers, Wahkeenah, WarthogDemon, Washburnmav, Wavehunter, Wavelength, Wayne Slam, Wbrice83186, WegianWarrior, Wehwalt, Welsh, Werdan7, Wetman, Wfaulk, WhisperToMe, Whispering, White Shadows, Whittlepedia, Whqttt, Wiggy!, Wik, WikiSvein, Wikipedian231, WildFan48, Will Beback, William Avery, WilliamH, Williamenicholls, Willking1979, Wimt, Winston365, Wisespider123, Wizardist, Wizzard2k, Wranid, Wtmitchell, Wv235, Ww2censor, X!, Xiahou, Xiner, Xiong, Xiong Chiamiov, Xnuala, Y2kcrazyjoker4, Yamamoto Ichiro, Yamla, Yanksox, Ye olde minr, Yellowtail, Yintan, Yosha, Yotsuya48, YourEyesOnly, Ytrewwasd, Yu61nus, Yuckfoo, Yvwv, ZX81, Zahid Abdassabur, Zanaq, Zandperl, Zapptastic, Zarcadia, Zarius, Zephyrad, Zlatan Ramić, Zonafan39, Zscout370, Zsinj, Zvika, Zyxw, Zzuuzz, Фальшивомонетчик, احمد, 3562 anonymous edits

Image Sources, Licenses and Contributors

Image:Tad Sendzimir.jpg *Source*: http://en.wikipedia.org/w/index.php?title=File:Tad_Sendzimir.jpg *License*: unknown *Contributors*: Brandmeister (old)

Image:POL COA Sędzimir.svg *Source*: http://en.wikipedia.org/w/index.php?title=File:POL_COA_Sędzimir.svg *License*: unknown *Contributors*: User:Avalokitesvara, User:Bastian

Image:Galvanized surface.jpg *Source*: http://en.wikipedia.org/w/index.php?title=File:Galvanized_surface.jpg *License*: unknown *Contributors*: Original uploader was Splarka at en.wikipedia

Image:Rust 1.jpeg *Source*: http://en.wikipedia.org/w/index.php?title=File:Rust_1.jpeg *License*: unknown *Contributors*: Original uploader was Peterlewis at en.wikipedia

File:redox reaction.png *Source*: http://en.wikipedia.org/w/index.php?title=File:Redox_reaction.png *License*: unknown *Contributors*: Original uploader was Bensaccount at en.wikipedia

File:Redox Halves.png *Source*: http://en.wikipedia.org/w/index.php?title=File:Redox_Halves.png *License*: unknown *Contributors*: Cameron Garnham

File:RustyChainEdit1.jpg *Source*: http://en.wikipedia.org/w/index.php?title=File:RustyChainEdit1.jpg *License*: unknown *Contributors*: Marlith. Original uploader was lamthedeus at en.wikipedia

File:Large bonfire.jpg *Source*: http://en.wikipedia.org/w/index.php?title=File:Large_bonfire.jpg *License*: unknown *Contributors*: User:Fir0002

Image:PyOx.JPG *Source*: http://en.wikipedia.org/w/index.php?title=File:PyOx.JPG *License*: unknown *Contributors*: User:Citypeek

File:Ascorbic acid structure.png *Source*: http://en.wikipedia.org/w/index.php?title=File:Ascorbic_acid_structure.png *License*: unknown *Contributors*: User:Jrockley

File:Dehydroascorbic acid.png *Source*: http://en.wikipedia.org/w/index.php?title=File:Dehydroascorbic_acid.png *License*: unknown *Contributors*: User:Yikrazuul

File:UraniumMineUtah.JPG *Source*: http://en.wikipedia.org/w/index.php?title=File:UraniumMineUtah.JPG *License*: unknown *Contributors*: Qfl247 (talk).

Image:Feuerverzinkte Oberfläche.jpg *Source*: http://en.wikipedia.org/w/index.php?title=File:Feuerverzinkte_Oberfläche.jpg *License*: unknown *Contributors*: TMg

Image:galvanizing01.jpg *Source*: http://en.wikipedia.org/w/index.php?title=File:Galvanizing01.jpg *License*: unknown *Contributors*: Ies, Phasmatisnox

Image:galvanizing02.JPG *Source*: http://en.wikipedia.org/w/index.php?title=File:Galvanizing02.JPG *License*: unknown *Contributors*: Ies, Phasmatisnox

Image:galvanizing03.JPG *Source*: http://en.wikipedia.org/w/index.php?title=File:Galvanizing03.JPG *License*: unknown *Contributors*: Ies, Phasmatisnox

Image:galvanizing04.jpg *Source*: http://en.wikipedia.org/w/index.php?title=File:Galvanizing04.jpg *License*: unknown *Contributors*: Ies, Phasmatisnox

Image:IOM3 logo.png *Source*: http://en.wikipedia.org/w/index.php?title=File:IOM3_logo.png *License*: unknown *Contributors*: Aaditya 7, Bill william compton

Image:POL COA Ostoja średniowieczna.svg *Source*: http://en.wikipedia.org/w/index.php?title=File:POL_COA_Ostoja_średniowieczna.svg *License*: unknown *Contributors*: User:Avalokitesvara, User:Bastian

Image:Herb Ostoja, Baranow Sandomierski.PNG *Source*: http://en.wikipedia.org/w/index.php?title=File:Herb_Ostoja,_Baranow_Sandomierski.PNG *License*: unknown *Contributors*: User:Gustavo Szwedowski de Korwin, User:Piotrus

Image:Ostoja herb.svg *Source*: http://en.wikipedia.org/w/index.php?title=File:Ostoja_herb.svg *License*: unknown *Contributors*: User:Bastianow

Image:POL COA Mikorski.svg *Source*: http://en.wikipedia.org/w/index.php?title=File:POL_COA_Mikorski.svg *License*: unknown *Contributors*: User:Avalokitesvara, User:Bastian, User:Waćpan

Image:POL COA Lachowicki- Czechowicz.svg *Source*: http://en.wikipedia.org/w/index.php?title=File:POL_COA_Lachowicki-_Czechowicz.svg *License*: unknown *Contributors*: User:Avalokitesvara, User:Bastianow, User:Waćpan

Image:POL COA Ostoja Ścibor.svg *Source*: http://en.wikipedia.org/w/index.php?title=File:POL_COA_Ostoja_Ścibor.svg *License*: unknown *Contributors*: User:Avalokitesvara, User:Bastian

Image:Ostoja run.jpg *Source*: http://en.wikipedia.org/w/index.php?title=File:Ostoja_run.jpg *License*: unknown *Contributors*: Camdan, Martin H., 1 anonymous edits

Image:Blason Ostoja - Sciborye (selon Gelre).svg *Source*: http://en.wikipedia.org/w/index.php?title=File:Blason_Ostoja_-_Sciborye_(selon_Gelre).svg *License*: unknown *Contributors*: User:Jimmy44

Image:Ostoja seal at thorn 1466.jpg *Source*: http://en.wikipedia.org/w/index.php?title=File:Ostoja_seal_at_thorn_1466.jpg *License*: unknown *Contributors*: none

Image:Beckov castle.jpg *Source*: http://en.wikipedia.org/w/index.php?title=File:Beckov_castle.jpg *License*: unknown *Contributors*: Peter Holly

Image:Krzyz-zaslugi Polska.png *Source*: http://en.wikipedia.org/w/index.php?title=File:Krzyz-zaslugi_Polska.png *License*: unknown *Contributors*: User:PawełMM, User:Pibwl

Image:POL Złoty Krzyż Zasługi BAR.svg *Source*: http://en.wikipedia.org/w/index.php?title=File:POL_Złoty_Krzyż_Zasługi_BAR.svg *License*: unknown *Contributors*: Alno, Ignasi, Mboro, Orem

Image:POL Srebrny Krzyż Zasługi BAR.svg *Source*: http://en.wikipedia.org/w/index.php?title=File:POL_Srebrny_Krzyż_Zasługi_BAR.svg *License*: unknown *Contributors*: Alno, Jarekt, Mboro, Orem

Image:POL Brązowy Krzyż Zasługi BAR.svg *Source*: http://en.wikipedia.org/w/index.php?title=File:POL_Brązowy_Krzyż_Zasługi_BAR.svg *License*: unknown *Contributors*: user:Orem

Image:Royal Swedish Academy of Sciences seal.png *Source*: http://en.wikipedia.org/w/index.php?title=File:Royal_Swedish_Academy_of_Sciences_seal.png *License*: unknown *Contributors*: Uppland

Image:KVA.JPG *Source*: http://en.wikipedia.org/w/index.php?title=File:KVA.JPG *License*: unknown *Contributors*: Dontworry, Hackspett, I99pema, Interpretix, LX, Prolineserver, Torben, Urbourbo

File:Vetenskapsakademiens handlingar 1750 front.jpg *Source*: http://en.wikipedia.org/w/index.php?title=File:Vetenskapsakademiens_handlingar_1750_front.jpg *License*: unknown *Contributors*: User:Tomas er

Image:Kungliga Vetenskapsakademien.jpg *Source*: http://en.wikipedia.org/w/index.php?title=File:Kungliga_Vetenskapsakademien.jpg *License*: unknown *Contributors*: User:Avjoska

Image:Statue of Liberty 7.jpg *Source*: http://en.wikipedia.org/w/index.php?title=File:Statue_of_Liberty_7.jpg *License*: unknown *Contributors*: User:Elcobbola

File:Flag of the United States.svg *Source*: http://en.wikipedia.org/w/index.php?title=File:Flag_of_the_United_States.svg *License*: unknown *Contributors*: Anomie

file:USA_New_York_City_location_map.svg *Source*: http://en.wikipedia.org/w/index.php?title=File:USA_New_York_City_location_map.svg *License*: unknown *Contributors*: User:Alexrk2

File:Red pog.svg *Source*: http://en.wikipedia.org/w/index.php?title=File:Red_pog.svg *License*: unknown *Contributors*: Anomie

File:U.S. Patent D11023.jpeg *Source*: http://en.wikipedia.org/w/index.php?title=File:U.S._Patent_D11023.jpeg *License*: unknown *Contributors*: Original uploader was Dpbsmith at en.wikipedia

File:LionDeBelfort.jpg *Source*: http://en.wikipedia.org/w/index.php?title=File:LionDeBelfort.jpg *License*: unknown *Contributors*: User:sanchezn

File:Indian princess and Columbia.jpg *Source*: http://en.wikipedia.org/w/index.php?title=File:Indian_princess_and_Columbia.jpg *License*: unknown *Contributors*: Connormah, Mangoman88, OptimumPx, Wehwalt

File:Freedom 1.jpg *Source*: http://en.wikipedia.org/w/index.php?title=File:Freedom_1.jpg *License*: unknown *Contributors*: Statue by Thomas Crawford (22 March 1814 – 10 October 1857) Photo by Architect of the Capitol

File:Collossal hand and torch. Bartholdi's statue of "Liberty.", from Robert N. Dennis collection of stereoscopic views.jpg *Source*: http://en.wikipedia.org/w/index.php?title=File:Collossal_hand_and_torch._Bartholdi's_statue_of_"Liberty."_from_Robert_N._Dennis_collection_of_stereoscopic_views.jpg *License*: unknown *Contributors*: Elcobbola

File:SOLparkParis.jpg *Source*: http://en.wikipedia.org/w/index.php?title=File:SOLparkParis.jpg *License*: unknown *Contributors*: Albert Fernique (born c. 1841, died 1898)

File:Pedestal for Bartholdi's Statue of Liberty.jpg *Source*: http://en.wikipedia.org/w/index.php?title=File:Pedestal_for_Bartholdi's_Statue_of_Liberty.jpg *License*: unknown *Contributors*: Original uploader was Hugh Manatee at en.wikipedia

File:EdwardMoran-UnveilingTheStatueofLiberty1886Large.jpg *Source*: http://en.wikipedia.org/w/index.php?title=File:EdwardMoran-UnveilingTheStatueofLiberty1886Large.jpg *License*: unknown *Contributors*: Edward Moran (1829-1901)

File:Statue of Liberty 1917 poster.jpg *Source*: http://en.wikipedia.org/w/index.php?title=File:Statue_of_Liberty_1917_poster.jpg *License*: unknown *Contributors*: Sackett & Wilhelms Corp. N.Y.

File:Liberty 1927.gif *Source*: http://en.wikipedia.org/w/index.php?title=File:Liberty_1927.gif *License*: unknown *Contributors*: Unknown photographer for Department of Agriculture

File:Nixon at Liberty Island.jpg *Source*: http://en.wikipedia.org/w/index.php?title=File:Nixon_at_Liberty_Island.jpg *License*: unknown *Contributors*: Jack Kightlinger, White House photographer

File:Nancy Reagan reopens Statue of Liberty 1986.jpg *Source*: http://en.wikipedia.org/w/index.php?title=File:Nancy_Reagan_reopens_Statue_of_Liberty_1986.jpg *License*: unknown *Contributors*: White House Photographic Office

File:National Park Service 9-11 Statue of Liberty and WTC fire.jpg *Source*: http://en.wikipedia.org/w/index.php?title=File:National_Park_Service_9-11_Statue_of_Liberty_and_WTC_fire.jpg *License*: unknown *Contributors*: National Park Service

File:Libertyferry.gif *Source*: http://en.wikipedia.org/w/index.php?title=File:Libertyferry.gif *License*: unknown *Contributors*: Arthur Tress, photographer for the Environmental Protection Agency

File:Emma Lazarus plaque.jpg *Source*: http://en.wikipedia.org/w/index.php?title=File:Emma_Lazarus_plaque.jpg *License*: unknown *Contributors*: melanzane1013

File:Statue of Liberty frontal 2.jpg *Source*: http://en.wikipedia.org/w/index.php?title=File:Statue_of_Liberty_frontal_2.jpg *License*: unknown *Contributors*: User:Dschwen

File:Las Vegas NY NY Hotel.jpg *Source*: http://en.wikipedia.org/w/index.php?title=File:Las_Vegas_NY_NY_Hotel.jpg *License*: unknown *Contributors*: user:Flicka

File:1971 airmail stamp C80.jpg *Source*: http://en.wikipedia.org/w/index.php?title=File:1971_airmail_stamp_C80.jpg *License*: unknown *Contributors*: usps

File:LineartPresRev.png *Source*: http://en.wikipedia.org/w/index.php?title=File:LineartPresRev.png *License*: unknown *Contributors*: Original uploader was Aido2002 at en.wikipedia Later versions were uploaded by ACupOfCoffee, Timrem at en.wikipedia.

Printed by Books on Demand GmbH, Norderstedt / Germany